Urban Sustainability

The Urban Sustainability Book Series is a valuable resource for sustainability and urban-related education and research. It offers an inter-disciplinary platform covering all four areas of practice, policy, education, research, and their nexus. The publications in this series are related to critical areas of sustainability, urban studies, planning, and urban geography.

This book series aims to put together cutting-edge research findings linked to the overarching field of urban sustainability. The scope and nature of the topic are broad and interdisciplinary and bring together various associated disciplines from sustainable development, environmental sciences, urbanism, etc. With many advanced research findings in the field, there is a need to put together various discussions and contributions on specific sustainability fields, covering a good range of topics on sustainable development, sustainable urbanism, and urban sustainability. Despite the broad range of issues, we note the importance of practical and policy-oriented directions, extending the literature and directions and pathways towards achieving urban sustainability.

The series will appeal to urbanists, geographers, planners, engineers, architects, governmental authorities, policymakers, researchers of all levels, and to all of those interested in a wide-ranging overview of urban sustainability and its associated fields. The series includes monographs and edited volumes, covering a range of topics under the urban sustainability topic, which can also be used for teaching materials.

Saeid Pourroostaei Ardakani · Ali Cheshmehzangi

Big Data Analytics for Smart Transport and Healthcare Systems

Springer

Saeid Pourroostaei Ardakani
School of Computer Science
University of Lincoln
Lincoln, Lincolnshire, UK

Ali Cheshmehzangi
Qingdao City University
Qingdao, Shandong, China

ISSN 2731-6483 ISSN 2731-6491 (electronic)
Urban Sustainability
ISBN 978-981-99-6619-6 ISBN 978-981-99-6620-2 (eBook)
https://doi.org/10.1007/978-981-99-6620-2

This Springer imprint is published by the registered company Springer Nature Singapore Pte Ltd.
The registered company address is: 152 Beach Road, #21-01/04 Gateway East, Singapore 189721, Singapore

Paper in this product is recyclable.

There are places we care about, and we hope they have chances for innovation, rejuvenation, advancement, and nurturing enrichment merely amplified in the right way. We collectively dedicate this to the Patria, where we feel connected (with) and wish to see prosperity, perpetual amelioration, and sturdiness in sustaining a brighter future. In other words, betterment astonishes rivals and yields enthusiasm kept highly above accomplishments. Knowledge-based evolutions erupt knowledgeable habitats, and smartness transforms environments yarning entrepreneurialism. Vision augments trust and nobility. Hence, the words and wishes are set for what could be a paradise beyond imagination. And with that, we also dedicate the book to the future of smartness, wishing it could become a rightly developed foundation for more proficient people, communities, and society. Innovation shall grow and shall lead to further success. We hope to witness that in our ephemeral lifetime.

Acknowledgements

We collectively acknowledge our research interns, team members, and external collaborators. All members have been extremely helpful in completing this book project. Despite the hardships, they have provided us with excellent support in collecting data, conducting research, and completing the tasks under each given project. Our research interns worked hard according to the assigned tasks and objectives. Our team members and external collaborators helped with reviews and gave us valuable feedback. We thank them all for their continuous support and hope to have longer-term relationships with them.

We reflect on our enduring friendship and invigorating collaboration, and we hope they last for many years, beyond anyone's imagination and beyond any superior control. The moments we spent together on research and everyday life are memorable days and nights, and we shall cherish them forever.

Ali Cheshmehzangi acknowledges the Ministry of Education, Culture, Sports, Science, and Technology (MEXT), the Japanese Government, and Hiroshima University, Japan.

About This Book

Big Data Analytics for Smart Transport and Healthcare Systems is the second volume of our work on big data analytics and smart urban systems. It aims to introduce big data solutions in urban sustainability applications—mainly smart transportation and healthcare systems. It focuses on machine learning techniques and data processing approaches which have the capacity to handle/process huge, live, and complex datasets in real-time transportation and healthcare applications. For this, several state-of-the-art data processing approaches including data pre-processing, classification, regression, and clustering are introduced, tested, and evaluated to highlight their benefits and constraints where data is sensitive, real-time, and/or semi-structured. The key benefits of this book are: (1) to introduce the principles of machine learning-enabled big data analysis in transportation and healthcare applications; (2) to present state-of-the-art data analysis solutions where data is sensitive, huge, real-time, and complex; and (3) to understand the principles of smart transportation and healthcare data analytics.

Prologue

Information is the oil of the twenty-first century, and analytics is the combustion engine

—Peter Sondergaard, Gartner, Inc.

In the age of smart everything, data—and particularly big data—dictates and determines many things. Cities have become excellent platforms of smart experimentation, sold and branded under the ideologies of innovation and technological advancement. Data-driven methods and solutions in urban systems have accelerated in popularity and applications, ranging from information-based analysis to AI integration and development.

The current situation/age/era reminds us of the 2002 Hollywood movie 'Minority Report,' in which the future was demonstrated based on data-based scenarios. The platforms where we are part of, from our personal devices to surveillance systems, suggest the pros and cons of such transitions in how we live and operate as (living) beings. Many existing and foreseeing threats suggest the slow pace of general acceptance, while we know the pathway is clearly set. Thus, we are spoon-fed at a gradual pace in order to avoid any societal shocks, disorders, and disruptions. In many ways, there are threats of control and not management, where top-down decisions could determine the livelihood and likelihood of our communities and, ultimately, our society. Nonetheless, we believe that scholars from both computer science and urbanism or urban studies should put forces together to not only better understand cities but to optimize them and make them better for people who live and work in them. If to let governments decide, then it will all be about control. If you let the city management teams and enterprises decide, it will all be about top-down management. If you let the general public decide, they will be clueless and inconsiderate about what can be done and how we can progress further. Thus, it is in the hands of enthusiastic and forward-thinking experts that we could rightfully advance our cities to become better living and working environments for all and not just for a few.

Data is floating in everyday activities and operations, and our job is to make sure it is captured and cleaned, selected and assessed, reviewed and managed, and finally integrated and utilized. To do so, we need big data specialists, data scientists, and urban experts. We need the support of multiple stakeholders to ensure our databases are accurate and not manipulated, our directions are correct and not forged, and our analysis is constructive and not deceptive. We have to break any potential misconceptions and misuse to ensure the future of our cities and communities is healthier, more resilient, and more sustainable.

Following our earlier book on *Big Data Analytics for Smart Urban Systems*, this second volume of our collaborative and interdisciplinary research work focuses on sector-based contributions, particularly related to smart transport and healthcare systems. The focus here helps us shed light on critical directions that are vital to sectoral and cross-sectoral research and practice. The two should be considered, or else we create more gaps than integration.

For us, criticality keeps intelligence near genuine big-data applications. Solutions transform analytical records, drive services, and transcend unique, nurturing, and noteworthy cities. For us, cities kick-start thresholds, i.e., hotspots to enrich prosperity and redemption to territorial instigations, innovation, and interactions. While we know we are far from achieving these, we would instead join forces against the unkind and unjust flows and move ahead toward an all-inclusive and human-centric future. With that, we hope to keep our networks and the healthiness of society as the priority so that we do not lose touch with humanity amid the many technological advancements and machine-based societal developments. We must keep a right mind to ensure ethics and humanity/society are maintained, sustained, and advanced. We trust there is light at the end, and we hope to see that light sooner before humanity gets into a darker and longer tunnel.

June 2023

Saeid Pourroostaei Ardakani
Ali Cheshmehzangi

Contents

1 **The Role of Big Data Analytics in Urban Systems: Review and Prospect for Smart Transport and Healthcare Systems** 1
1 Introduction 1
2 Big Data Analytics in Urban Systems 2
2.1 Development Process of Big Data Analytics 3
2.2 Application of Big Data Analytics 4
2.3 Security and Detection 4
2.4 Energy Consumption Management 5
2.5 Governance and Planning 6
2.6 Health Monitoring and Response 6
2.7 Transportation and Flows 6
3 Challenges of Big Data Analytics in Cities 7
4 Prospects of Urban Big Data Analytics 8
4.1 Review and Prospect for Smart Transport and Healthcare Systems 9
5 Conclusions 9
6 Summary 10
Appendix 15
References 24

Part I Smart Transport

2 **Big Data Analysis for an Optimised Classification for Flight Status: Prediction Analysis Using Machine Learning Classifiers** 29
1 Introduction 29
2 Literature Review 31
2.1 Machine Learning for Flight Predictions 32
3 Methodology 33
4 Result and Discussion 37
5 Conclusion 40

References ... 42

3 **On-Board Unit Freight Transport Data Analysis and Prediction: Big Data Analysis for Data Pre-processing and Result Accuracy** ... 45
1 Introduction ... 45
2 Literature Review ... 46
3 Methodology ... 49
3.1 Data Preprocessing ... 49
3.2 Time Series Prediction Models ... 50
4 Results and Discussions ... 53
4.1 Data Exploration and Visualisation ... 53
4.2 Contributions in Data Preprocessing ... 56
4.3 Comparison of LSTM and LSTM + FCN Model ... 56
5 Conclusion ... 59
References ... 60

4 **Data-Driven Multi-target Prediction Analysis for Driving Pattern Recognition: A Machine Learning Approach to Enhance Prediction Accuracy** ... 63
1 Introduction ... 63
2 Literature Review ... 66
3 Methodology ... 68
3.1 Data Pre-processing ... 68
3.2 Multi-target Prediction Model ... 72
3.3 Clustering ... 73
4 Result and Discussion ... 74
4.1 Performance Evaluation Metrics ... 75
4.2 The Model Evaluation ... 75
4.3 Model Training Delay ... 77
5 Conclusion and Future Work ... 77
References ... 78

5 **A Predictive Data Analysis for Traffic Accidents: Real-Time Data Use for Mobility Improvement and Accident Reduction** ... 81
1 Introduction ... 81
2 Literature Review ... 82
3 Methodology ... 84
3.1 Data Preprocessing ... 84
3.2 Information Correlation Calculation ... 86
4 Result and Discussion ... 88
4.1 Data Analysis ... 88
4.2 Correlation Calculation ... 93
4.3 Evaluation ... 93
5 Conclusion ... 93
References ... 97

Part II Smart Healthcare

6 Healthcare Infrastructure Development and Pandemic Prevention: An Optimal Model for Healthcare Investment Using Big Data ... 103
1 Introduction ... 103
2 Literature Review ... 104
3 Methodology ... 107
3.1 Data Processing ... 107
3.2 Linear Regression ... 109
3.3 SVR ... 109
3.4 KNN ... 109
3.5 Decision Tree ... 110
4 Results and Discussions ... 110
4.1 Linear Regression Analysis ... 110
4.2 SVR ... 112
4.3 KNN ... 115
4.4 Decision Tree ... 116
4.5 Discussions ... 116
5 Conclusions ... 117
References ... 118

7 Big Data for Social Media Analysis During the COVID-19 Pandemic: An Emotion Analysis Based on Influences from Social Networks ... 121
1 Introduction ... 121
2 Literature Review ... 123
2.1 Hot Topic Searches: An Analysis of Information Spread During the COVID-19 Pandemic ... 123
2.2 Social Media Sentimental Analysis Abroad ... 124
2.3 Social Media Sentimental Analysis in China ... 125
2.4 Public Opinions Toward COVID-19 Lockdown ... 126
2.5 Text Emotion Analysis ... 126
3 Methodology ... 128
3.1 Data Collection ... 128
3.2 Data Pre-processing ... 129
3.3 Data Analysis and Visualisation ... 131
3.4 Machine Learning ... 132
4 Results and Discussion ... 132
5 Conclusions ... 137
References ... 138

8 Big Data-Enabled Time Series Analysis for Climate Change Analysis in Brazil: An Artificial Neural Network Machine Learning Model ... 141
1 Introduction ... 141

2 Related Work 142
2.1 PCA 142
2.2 Cross-Validation 143
2.3 Time Series Model 144
2.4 Multi-layer Perceptron 145
3 Methodology 146
3.1 Dimension Reduction 146
3.2 Time Series Model 146
4 Results and Discussion 150
4.1 Results of PCA 150
4.2 Results of Model 151
4.3 Results Analysis 153
5 Conclusions and Future Work 154
References 154

9 Optimized Clustering Model for Healthcare Sentiments on Twitter: A Big Data Analysis Approach 157
1 Introduction: Social Media and Public View on Health-Related Matters 157
2 Literature Review on Sentiment Analysis in Healthcare 159
2.1 Keywords Extraction Methods 160
2.2 Sentiment Analysis Clustering Models 161
3 Methodology 162
3.1 Feature Extraction Algorithm 162
3.2 Clustering Algorithms 163
4 Experiments and Results 165
4.1 Dataset 165
4.2 Experiment and Configuration 165
4.3 Evaluation Metrics 166
4.4 Clustering Results and Discussion 166
5 Conclusions 170
References 171

10 Big Data Analytics and the Future of Smart Transport and Healthcare Systems 175
1 A Brief Reflection on Big Data Analytics and Smart Transport and Healthcare Systems 175
2 Sectoral Contributions of the Book 177
3 Concluding Remarks: A Summary of Lessons Learnt for Future Research 179
3.1 On 'Smart Transport' 179
3.2 On 'Smart Healthcare' 180
References 182

Index 185

About the Authors

Saeid Pourroostaei Ardakani currently works as Senior Lecturer in Computer Science at the University of Lincoln, UK. He is also an associated academic member of Lincoln Centre for Autonomous Systems (L-CAS) and has formerly worked at the University of Nottingham (UNNC) and Allameh Tabatabai University (ATU), as an assistant professor in Computer Science, member of the Next Generation Internet of Everything Laboratory (NGIoE) and Artificial Intelligent Optimisation Research group, and head of ATU-ICT center. He received his Ph.D. in Computer Science from the University of Bath focusing on data aggregation routing in wireless sensor networks. Saeid's research and teaching expertise centers on smart and adaptive computing and/or communication solutions to build collaborative/federated (sensory/feedback) systems in Internet of things (IoT) applications and cloud environments. He is also interested in (ML-enabled) big data processing and analysis applications. Saeid has published more than 60 scholarly articles in reputed international journals and peer-reviewed conferences.

Ali Cheshmehzangi is the world's top 2% field leader, recognized by Stanford University. He has recently taken a senior leadership and management role at Qingdao City University (QCU), where he is Professor in Urban Planning, Director of the Center for Innovation in Teaching, Learning, and Research, and Advisor to the school's international communications. Over 11 years at his previous institute, Ali was Full Professor in Architecture and Urban Design, Head of the Department of Architecture and Built Environment, Founding Director of the Urban Innovation Lab, Director of Center for Sustainable Energy Technologies, and Director of Digital Design Lab. He was Visiting Professor and now Research Associate of the Network for Education and Research on Peace and Sustainability (NERPS) at Hiroshima University, Japan. Ali is globally known for his research on 'urban sustainability.' So far, Ali has published over 300 journal papers, articles, conference papers, book chapters, and reports. To date, he has 15 other published books.

Chapter 1
The Role of Big Data Analytics in Urban Systems: Review and Prospect for Smart Transport and Healthcare Systems

Abstract This introduction chapter provides an overview of the idea, aim, and objectives of the book. It delves into the importance of Big Data as a tool in the current information age. The chapter starts with an overview of Big Data analytics for urban systems and follows the discussions from the sector-based perspectives. It then explores Big Data Applications (BDA) in two key areas of smart transportation and healthcare, particularly in cities and as part of (smart) urban systems. After providing a review of prospects about Big Data analytics in these two sectors, the chapter introduces the book structure and all case study chapters. This chapter shares an overall picture of the book to readers before we delve into global case study examples.

Keywords Big Data Analytics · Smart Cities · Transportation · Digital Healthcare · Urban Systems

1 Introduction

In the information age, big data has become an important tool for us to understand and improve the world, especially in urban planning and design (Pourroostaei Ardakani and Cheshmehzangi 2023), and the application of big data has triggered a revolution (Gandomi and Haider 2015; Cheshmehzangi et al. 2021; Cheshmehzangi 2022a, 2022b). We are now facing a fast-approaching paradigm shift to effectively mine and record relevant data with massive amounts of data collected automatically or voluntarily in the cloud.

Big Data refers to data sets that are larger than the collection, storage, management, and analysis capabilities of classical database systems (Kaisler et al. 2013; Wu et al. 2014; Pourroostaei Ardakani and Cheshmehzangi 2023). By collecting and analysing vast amounts of data, stakeholders can better understand the needs and problems of cities, leading to smarter and more liveable cities (Zikopoulos et al. 2012; Yamagata et al. 2020). In this opening chapter, the authors explore how big data is being

S. Pourroostaei Ardakani and A. Cheshmehzangi, *Big Data Analytics for Smart Transport and Healthcare Systems*, Urban Sustainability,
https://doi.org/10.1007/978-981-99-6620-2_1

applied to urban systems and how cities in the future can leverage big data analytics. In particular, we focus on smart transport and healthcare systems.

2 Big Data Analytics in Urban Systems

The process of identifying trends, patterns, and linkages in vast volumes of raw data in order to make data-driven choices is referred to as big data analytics (BDA). It has emerged as a critical area of research in recent years, driven by the exponential growth in data volume, velocity, and variety (Pourroostaei Ardakani and Cheshmehzangi 2023). The amount of data available to urban planners, municipalities, and the public is exploding. The challenge now is moving from generating data to organising and extracting data meaningfully (Cheshmehzangi et al. 2022, 2023). In the past, data was carefully collected, classified, structured, validated, and stored by trained experts, but now millions of people and billions of sensors are constantly generating and supplementing data and are constantly being classified, structured, and validated to form cloud-based data storage (Shvachko et al. 2010). Chen et al. (2014) defined big data as "*data that exceeds the processing capacity of conventional database systems due to its large size, high frequency of generation or complexity*". In addition to the available data that we actively acquire, the design, sensing, and measurement technologies that will be applied to specific engineering projects in the future can be autonomously collected and supplemented with other data from smartphone location, traffic, and other data voluntarily published by collectors.

As shown in Fig. 1.1, the word cloud in big data analytics and urban systems is quite diverse but focuses more on technologies, systems, frameworks, and services (such as ICT services). The diverse correlation between these two keywords suggests the wide range of BDA applications in cities and urban systems. Hence, by following our earlier work on methodological contributions of BDA to smart urban systems (Pourroostaei Ardakani and Cheshmehzangi 2023), we delve into particular smart systems and focus more on sector-based contributions. In this regard, we explore case study examples of two main sectors in smart urban systems, i.e., 'smart transport' and 'smart healthcare'. Here, we provide definitions of both systems:

Smart Transport or Smart Transportation

There is a wide range of benefits of smart transportation in cities. The US Department of Transportation refers to smart transportation systems as a sort of 'Intelligent Transportation Systems (ITS)', which "*apply a variety of technologies to monitor, evaluate, and manage transportation systems to enhance efficiency and safety*" (see Mazur 2020). According to Mazur (2020), three key concepts are suggested for smart transport systems, including management, efficiency, and safety, meaning that "*smart transportation uses new and emerging technologies to make moving around*

Fig. 1.1 Word cloud on big data analytics and urban systems. Drawn by the authors, the data is derived from the keywords of theme papers in the Web of Science database

a city more convenient, more cost effective (for both the city and the individual), and safer".

Smart Healthcare

In the literature, there are many studies about smart healthcare, and each of them defines the term differently. Much of the focus on smart healthcare refers to smart technologies that are applied to transform healthcare systems and services. According to the World Economic Forum (2021), four main areas of artificial intelligence (AI), remote technologies, data-driven healthcare, and smart hospital management are prioritised in this sector. In a way, we can say the revolutionary directions in smart healthcare are technology-based. In contrast, integration and innovation within healthcare remain to be fundamental to applying such technologies in practice (Pourroostaei Ardakani 2017). Thus, the shift to digital innovation and bringing in smart technologies into this sector is inevitable.

2.1 Development Process of Big Data Analytics

The BDA is a rapidly growing field that deals with the processing and analysis of large and complex datasets. The term "big data" refers to datasets that are too large, diverse, and dynamic for traditional data processing systems to handle effectively. The ability to extract insights, patterns, and knowledge from these vast datasets has opened up new opportunities for businesses, governments, and researchers (Manyika et al. 2011a). BDA involves the use of advanced tools and techniques from computer science, statistics, and mathematics to extract insights, patterns, and knowledge from these vast datasets (Pourroostaei Ardakani et al. 2021b; Nan et al. 2023). Some critical technology for BDA contains NoSQL databases (Han et al. 2011) and Hadoop (Shvachko et al. 2010), an open-source software framework for distributed storage and processing of large datasets. In addition to these technical innovations, there is growing interest in the development of ethical and responsible practices for BDA. Yang et al. (2017) argue that great power comes with great responsibility, and today's massive amount of data necessitates careful consideration of privacy, security, and ethical issues.

The development of BDA has been fuelled by the exponential growth in data volume, velocity, and variety, driven by the proliferation of digital technologies and the internet (Fan and Bifet 2013). The ability to collect, store, and analyse massive amounts of data has opened up new opportunities for businesses, governments, and researchers to gain valuable insights into various aspects of human behaviour, socio-economic trends, and environmental phenomena. Some of the key applications of BDA include customer behaviour analysis, predictive modelling, fraud detection, risk management, supply chain optimization, and personalized medicine (Davenport 2014; Manyika et al. 2011b). The field of BDA has also spawned related areas such as machine learning, artificial intelligence, and data science.

As the field of BDA continues to evolve, researchers and practitioners need to stay up to date with the latest developments and best practices. In the coming decade, the capacity to derive insights from large data will become a major competitive edge (Marr 2015; Provost and Fawcett 2013), making it essential for organizations to invest in the necessary skills and technologies. In summary, BDA is a rapidly growing field with far-reaching implications for businesses, governments, and society as a whole. While the challenges are significant, the potential benefits are equally compelling, making it an exciting area of research and innovation.

2.2 Application of Big Data Analytics

The BDA can study and impact urban systems in the following five ways (see Fig. 1.2):

2.3 Security and Detection

The most fundamental human need is to live in a secure environment. Security concerns in heavily networked cities extend to other critical sectors, such as energy and data security. Designing robust centralized or decentralized networks and associated elements is a significant challenge for future cities. For example, analysing crime patterns and trends might assist law enforcement organisations in better anticipating and responding to criminal activities (Mohler et al. 2011; Feng et al. 2019). Avazov et al. (2021) developed a deep learning-based framework for fire detection in urban areas. BDA can play a crucial role in enhancing the security of urban systems by providing real-time monitoring and detection of potential threats. For instance, city surveillance systems equipped with video cameras and sensors can capture a large

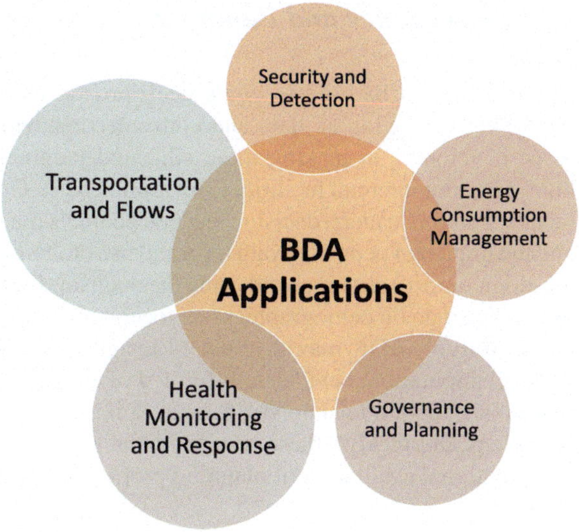

Fig. 1.2 Summary of BDA applications in five ways, with two areas of healthcare and transportation as our focus. *Source* The Authors

volume of data, which can be analysed using big data techniques to identify suspicious activities and behaviours. Furthermore, predictive analytics can help anticipate potential security risks, allowing authorities to take preventive measures before any incident occurs.

2.4 Energy Consumption Management

The BDA can be used to enhance energy use in urban systems, lowering costs while also decreasing environmental effects (Pham et al. 2021). By evaluating data from smart meters, weather predictions, and other sources, patterns, and trends in energy consumption may be identified, which can influence decisions about when and how to utilise energy more efficiently (Zhou et al. 2016; Al-Ali et al. 2017; Pourroostaei Ardakani and Cheshmehzangi 2023). The data analysis from smart meters, for example, can assist utilities in identifying regions of high energy use and developing methods to minimize energy consumption (Zhang et al. 2018; Fathi et al. 2020). The shift from a centralised energy supply to a coexistence of centralised and decentralised modes of energy production and supply will create massive amounts of data that must form the basis for future design and its interaction with spatial forms at the urban scale and building scale.

2.5 Governance and Planning

The BDA could be used to guide decisions on urban planning and development. Analysing demographic data and consumer behaviour trends, for instance, might assist developers in making more educated judgments about where to build new housing or commercial facilities (Townsend 2013). Governance institutions, election results, opinion leaders, decision leaders, people's reactions to government decisions, and the impact of people's evaluation and use of their environment on urban form and development will all serve as filtering mechanisms for big data and urban planning. By examining data on demographics, socioeconomic characteristics, and environmental circumstances, BDA may give useful insights for urban government and development. Computational evidence-informed planning is based on quantitative methods that allow generalizing knowledge and regularities from the analysis of existing cities or regions and making these results available for future planning. Respectively for computational evaluation of planning proposals based on past studies.

2.6 Health Monitoring and Response

The BDA can help in monitoring and responding to health-related issues in urban areas by analysing data on public health, diseases, and environmental factors (Khanra et al. 2020). For instance, evaluating social media data can assist in identifying disease outbreaks or places where there is a higher risk of infection (Kar et al. 2022). This data may subsequently be utilised to create targeted public health interventions (Kruse et al. 2016). Pramanik et al. (2017) proposed that the combination of big data and intelligent systems can accelerate the prospects of the healthcare industry.

2.7 Transportation and Flows

The BDA can be used to optimize transportation systems to reduce traffic congestion and improve mobility. The volume of timely updated traffic data is rapidly expanding and will form the basis for analysis and design inputs. Research on the close relationship between transportation and planning will provide an important result orientation for planning and design. Real-time data from GPS devices and traffic cameras, for example, can assist in detecting regions of congestion and help planners make better-educated decisions regarding traffic flow and public transport routing (Zhu et al. 2018; Pourroostaei Ardakani et al. 2021a). Urban flow analysis and its integration with design, big data mining, and the application of the results of logistics, transportation, economics, and material flow analysis all contribute to visual high-end dynamic design response.

3 Challenges of Big Data Analytics in Cities

However, the field of BDA also faces significant challenges, including issues of privacy, security, and governance. Kitchin (2014) examines the impact of big data on knowledge production and argues that it represents a paradigm shift in the way we understand and analyze data. Al Nuaimi et al. (2015) discuss the use of big data in urban planning, policymaking, and service delivery, highlighting the challenges and opportunities associated with its application. Kitchin and McArdle (2016) discuss the ontological characteristics of big data and highlights the challenges associated with its analysis and interpretation.

The specific challenges are reflected in the following aspects (see Fig. 1.3):

- ***Data Quality***: Big data is frequently unstructured, incomplete, or inconsistent, making correct analysis challenging.
- ***Data Privacy and Security***: As more data is gathered, shared, and analyzed, guaranteeing the privacy and security of sensitive information becomes more difficult.
- ***Interoperability***: Integrating data from diverse sources is difficult due to differences in systems and data formats.
- ***Human Factor***: Due to the complexity of big data analytics, trained human resources and effective communication amongst diverse stakeholders are required.
- ***Decision-Making***: Analyzing and interpreting data takes a substantial amount of time and effort, resulting in decision-making delays.
- ***Infrastructure***: Adequate technological infrastructure, such as high-speed networks and storage, is required for big data processing and analysis.

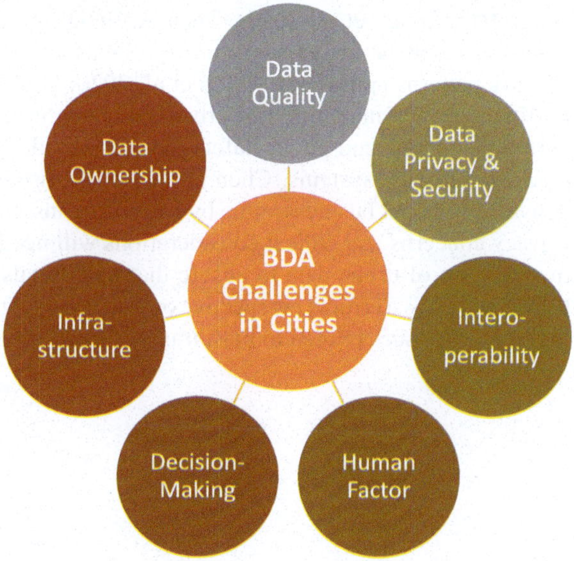

Fig. 1.3 Summary of BDA challenges in cities. *Source* The Authors

- ***Data Ownership***: Determining who owns the data generated by cities and how it can be used ethically and legally is a significant challenge.

As Kaisler et al. (2013) note, concerns regarding privacy, confidentiality, intellectual property rights, and accountability for erroneous or biased findings have arisen as a result of the increased dependence on big data (also see: Cheshmehzangi et al. 2023). To address these issues, academics and practitioners are creating new tools and strategies for organizing and interpreting large amounts of data. NoSQL databases, which provide flexible data structures and scalability for processing vast volumes of unstructured data (Han et al. 2014), are one significant area of study.

4 Prospects of Urban Big Data Analytics

Compared with traditional data, big data has the advantages of massive data scale, multi-source data types, dynamic spatial–temporal attributes, low-value density, and fast processing speed. The characteristics of big data analysis, such as multi-source, human-oriented, and spatial–temporal attributes, are tightly coupled with the essential attributes of urban systems, which makes its impact on urban planning more significant. Using big data thinking, methods and technical means, urban problems can be analysed and studied accurately, quantitatively, and meticulously, and the security scheme is more scientific. The layout of urban infrastructure and public service facilities will be more reasonable and efficient. In terms of data acquisition methods, spatial–temporal scales, and core values, it has a very important impact on urban planning development and transformation. First, increased Internet of Things (IoT) device and sensor usage will deliver more real-time data streams and enhance data quality (Karkouch et al. 2016). Then Artificial intelligence and machine learning algorithms will be increasingly used to analyse large amounts of urban data and identify patterns and trends (Ullah et al. 2020; Pourroostaei Ardakani et al. 2023). Furthermore, the integration of urban data from many sources (for example, transportation, energy, and public safety) would provide a more thorough knowledge of cities as complex systems. Chen et al. (2019) discuss how big data analytics can deliver significant business value by leveraging insights from large datasets. Finally, Privacy concerns and ethical considerations will need to be addressed in the collection and use of urban data to ensure that individuals' rights are protected. Kitchin (2017) provides a critical perspective on the role of algorithms in urban data analysis, highlighting issues related to bias, transparency, and accountability.

4.1 *Review and Prospect for Smart Transport and Healthcare Systems*

As mentioned earlier, two out of the five main areas for the application of BDA in cities are 'smart transport' and 'smart healthcare' systems. Correlated with the previous volume focused on smart urban systems, this book aims to put together a comprehensive set of case study examples in smart transport and healthcare systems. Thus, the book aims to introduce big data analytics applications and solutions in these two major sectors in cities (i.e., smart transportation and healthcare systems). For this, several state-of-the-art data processing approaches, including data pre-processing, classification, regression, and clustering, are introduced, tested, and evaluated to highlight their benefits and constraints where data is sensitive, real-time, and/or semi-structured. Similar to our previous volume on smart urban systems, this book's ultimate objectives or goals are to introduce the principles of machine learning-enabled big data analysis in smart transport and healthcare systems. Through global case study examples, we will present state-of-the-art data analysis solutions where data is sensitive, huge, real-time, and complex. Lastly, the case study examples would help us develop a holistic review and prospect for smart transport and healthcare systems. In doing so, our focus would be to provide knowledge about various solutions to understand the principles of smart transportation and healthcare data analytics. We believe these two sectors are primary to smart transformations and transitions; hence ICT-driven and data-driven approaches would play a major part in making such important urban systems smarter. In light of this, the book is divided into two parts of transportation and healthcare, each with four global case study examples.

5 Conclusions

The impact of big data on cities is huge and far-reaching. On the one hand, it has a huge impact on traditional urban planning concepts and methods, strengthens the control and decision-making power of planning on urban development, and provides new directions and technical means for the transformation of existing urban planning. On the other hand, it is of great significance to reconstruct and enrich the urban system and make it intelligent and low-carbon. The BDA provides an exponentially growing number of sources for high-quality decision-making, both now and in the future. Therefore, it becomes crucial to keep the data interactive to derive a more objective view.

In summary, by collecting and analysing large amounts of data, BDA can better understand the needs and problems of cities to design smarter and more liveable cities. For example, we may better comprehend a city's traffic circumstances and develop a more efficient transportation system by examining traffic data. We can comprehend cities' environmental concerns and develop more ecologically friendly cities by analysing environmental data. The urban system provides a platform for the application of BDA. In the information city, data is widely collected and used. For example, a smart home system can collect and analyse the energy usage data of the home, thus helping the home to save energy. Intelligent transportation systems can collect and analyse traffic data to help cities improve traffic conditions.

However, the relationship between BDA and urban systems is not just about supporting each other. In some cases, they may also generate conflicts. For example, the collection and use of big data may raise privacy and security issues, which may affect the health of cities. Therefore, we need to find a balance between the application of BDA and the construction of cities to ensure that we can enjoy the benefits of BDA while protecting our privacy and security to build smarter and more liveable cities.

6 Summary

This book contributes to applications and solutions of big data analytics in smart transportation and healthcare systems. In the following eight chapters, we explore case study examples and various big data analytics solutions. These sector-based applications and solutions are highlighted as major methods in the field, with great potential or scope for further development and integration in other smart urban systems. While we mainly focus on sectorial contributions, global case study examples are used to present, visualise, and summarise big data analytics methods in smart transportation and healthcare systems. To follow these debates, the book is structured into two parts, each dedicated to the specific urban systems of transport/transportation and healthcare. Chaps. 2–9 will provide some case study examples of BDA in smart transport (Part 1, Chaps. 2–5) and smart healthcare (Part 2, Chaps. 6–9). The following eight book chapters are summarised below.

Part 1: Smart Transport (with four case study chapters)

Chapter 2: Big Data Analysis for an Optimised Classification for Flight Status: Prediction Analysis using Machine Learning Classifiers

Abstract

Accurate flight delay forecasting is critical for establishing a more efficient airline industry. A smart system in place would help reducing the negative impacts of such delays. Machine learning-enabled Big data solutions have been widely utilized in recent studies to anticipate aircraft delays. They need a data pre-processing to understand and grasp the relevance of each data attribute. The results of data attribute relevance are used to filter out data deemed relevant to aircraft delays and eliminate the data that was redundant and unsuitable for analysis. This chapter trains linear and polynomial regression models to predict the delay time of a flight. The data analysis algorithm runs on a well-known dataset, which comprises flight data from more than ten US airlines from 2009 to 2019. The result indicates that 97.04% of the predicted result has a difference of fewer than 15 min between the actual value.

Keywords: Flight delay prediction; Polynomial regression; Linear regression; Data pre-processing; Pyspark.

Chapter 3: On-Board Unit Freight Transport Data Analysis and Prediction: Big Data Analysis for Data Pre-processing and Result Accuracy

Abstract

Traffic prediction is a complex urban problem that needs to collect and analyse big data and train accurate machine learning models. Solving it will contribute to society's efficient production and development, particularly in optimising urban systems. This chapter aims to build a precise model to predict the number of freight vehicles in future timestamps based on historical data provided On-Board Unit (OBU) datasets in Belgium's road networks. We contributed to two novel solutions to solve the time series prediction task. The first contribution is preprocessing the data using SparkSQL and generating nine features from the prediction timestamps. The second is LSTM and LSTM + FCN deep learning models with tuned parameters and training with the newly generated features. The results of our LSTM model are more accurate than those of other models on this dataset that we have already examined, reaching an accuracy of 99.89%. The preprocessing stage is proven to be vital for the performance of the LSTM model. As a result, models using our preprocessed nine features are much more accurate on this freight transportation prediction task. Discussions are also raised to understand better freight transport prediction tasks from the data and results points of view.

Keywords: On-Board Unit; Time series traffic prediction; Freight transport; Urban system; LSTM.

Chapter 4: Data-driven Multi-target Prediction Analysis for Driving Pattern Recognition: A Machine Learning Approach to enhance Prediction Accuracy

Abstract

The driving pattern is critical in life quality enhancement, road traffic minimisation, and transportation risk reduction. It includes several parameters, such as the car's speed, traffic, weather, and road status. This chapter investigates the correlations between driving attributes and proposes a multi-target prediction model to recognise driving patterns. For this, a pre-processing data approach, including Pearson's Correlation Coefficient, Quantile Discretisation, and Hybrid Data Resampling, is used to reduce data feature dimensions and remove meaningless variables. The Random Forest technique predicts four targets, including vehicle speed, rain intensity, driver's well-being, and driver's rush. At the same time, the K-Nearest Neighbours algorithm (KNN) groups the prediction results and forms the driving patterns. The performance of the proposed multi-target prediction model is compared with four classifiers, including Multilayer Perceptron, Decision Tree, Multinomial Logistics Regression, and LR one vs. rest. According to the results, the Random Forest model outperforms the benchmarks regarding prediction accuracy. The findings of this chapter help optimise prediction accuracy that could then be used for urban transportation system optimisation.

Keywords: Driving pattern recognition; Multi-target prediction; Road traffic; KNN; Pearson correlation.

Chapter 5: A Predictive Data Analysis for Traffic Accidents: Real-time Data use for Mobility Improvement and Accident Reduction

Abstract

Road traffic accidents are a significant component of human and economic losses. In order to implement effective policies or carry out safe traffic travel and improve the road system's safety, it is essential to identify the data patterns in the dataset for extracting the main influencing features associated with traffic accidents. The analysis of traffic accidents is complex as they affect each other and are also affected by many other factors. We used the US Accidents dataset, covering 2.8 million accident records in 49 states in the United States from February 2016 to December 2021. Maximum Relevance (MR) is used to get feature relevance by obtaining Mutual Information (MI) of features. The traffic accident predictive models are expected to be more accurate by using feature relevance to assist people in making real-time transportation decisions to improve mobility and reduce accidents. The findings help improve urban transportation network and systems at multiple spatial levels.

Keywords: Unsupervised Learning; Traffic Accident; Road system; Safety; Feature Relevance; Mutual Information.

Part 2: Smart Healthcare (with four case study chapters)

Chapter 6: Healthcare Infrastructure Development and Pandemic Prevention: An Optimal Model for Healthcare Investment using Big Data.

Abstract

This chapter proposes an optimal model for predicting the minimum healthcare investment required for pandemic prevention. It focuses on the COVID-19 pandemic as a recent example to further analyse the role of healthcare infrastructure development. It highlights the relationship between the control of COVID-19 and infrastructure development using linear regression, SVR, KNN, and decision tree models. By analysing the impact of each feature on the results, we select appropriate attributes and data processing methods. The findings of this study demonstrate that KNN is the best model with the highest training score of 0.655, while the training result of the decision tree is the worst, and the score is no more than 0.5. Lastly, the study highlights how big data could be used to improve the availability and development of urban critical infrastructures, such as healthcare infrastructure. An optimal model is suggested as part of the conclusion of this study.

Keywords: Healthcare; Infrastructure development; COVID-19; Big data Analysis; Machine Learning.

Chapter 7: Big Data for Social Media Analysis during the COVID-19 Pandemic: An Emotion Analysis based on Influences from Social Networks

Abstract

Since the beginning of the COVID-19 outbreak, the situation has soon become the most popular topic across social networks worldwide. For almost three years, news and social media were covered with information, updates, and daily/regular reports on the COVID-19 pandemic. When face-to-face activities were restricted, people started to use social networks more than before. The situation led to the empowerment of digital media, such as social media and networks. These platforms became the leading online social hubs for people to express their feelings and record their daily lives during the pandemic. In this study, we analysed this critical societal change as a major topic, mainly related to the pandemic on Weibo, which is the most popular social media platform in China. The study is focused on one particular context allowing in-depth data analysis related to China's societal impact due to the COVID-19 pandemic. The study also investigates the relationship between COVID-19 trends or topics and public sentiments on social networks. Machine learning is used to verify the correlation between emotion on social media and the COVID-19 pandemic trends. The study concludes with a prediction model using big data for public sentiments on social media.

Keywords: COVID-19; Emotion; Social networks; Weibo; Text mining.

Chapter 8: Big Data-enabled Time Series analysis for Climate Change Analysis in Brazil: An Artificial Neural Network Machine Learning Model.

Abstract

Climate data is an essential kind of data for humans in the world. Improving the ability to forecast the climate will contribute to the development of many industries, such as agriculture and shipping. In this chapter, we use the climate data in Brazil from 2000 to 2020. Attributes of the data mainly are date and time, temperature, precipitation, wind speed, and the province in which these data are measured. This study aims to classify these climate data and analyse the changing climate trends in the same province. An artificial neural network is established as the model in this project to implement this objective. The performance shows that this model can complete this classification task.

Keywords: Time series Analysis; Classification; Climate; Artificial neural network; Brazil.

Chapter 9: Optimized Clustering Model for Healthcare Sentiments on Twitter: A Big Data Analysis Approach.

Abstract

Social media, such as Twitter, typically stores a large amount of user-generated content regarding different aspects of society. These contents include social events, e-commerce products, healthcare, etc. This chapter proposes a best-fitted clustering method to classify sentiment samples related to healthcare topics. Thus, we examine other clustering models with keyword extraction methods on the real healthcare datasets collected from Twitter. The experiment results indicate that self- organized map model with the TF-IDF extraction method can achieve the best clustering accuracy. Moreover, the optimized model can have great potential to handle large-scale data in real practice.

Keywords: Index Terms; TF-IDF; Healthcare; Sentiments analysis; Clustering.

In addition, at the end of this chapter, we provide examples of smart transportation and healthcare systems (see Boxes 1.1 to 1.10).

Appendix

See Boxes 1.1, 1.2, 1.3, 1.4, 1.5, 1.6, 1.7, 1.8, 1.9 and 1.10.

Box 1.1 Examples of 'Smart Transportation' System.

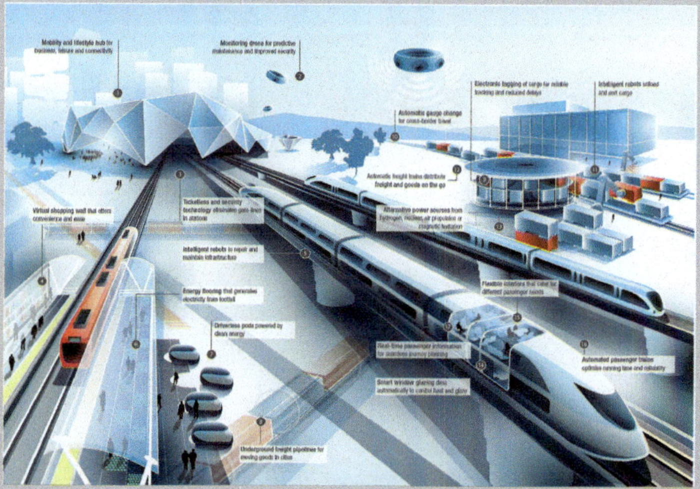

2014 Report

Arup's Vision of the Future of the Rail

> A new report by Arup reveals a vision of the future of rail travel in light of trends such as urban population growth, climate change and emerging technologies. Future of Rail 2050 foresees predictive maintenance of rail lines by robot drones, driverless trains travelling safely at high speed, freight delivered automatically to its destination, and smart technology able to interface with mobile and wearable devices to improve passenger experience and enable ticketless travel.

In three areas of: (1) Convenience (a reliable network), (2) Conurbations (integrated transport systems), and (3) Connectivity (plugging into journeys).

Available from: https://www.bdcnetwork.com/arups-vision-future-rail-driverless-trains-maintenance-drones-and-automatic-freight-delivery.

Box 1.2 Examples of 'Smart Transportation' System

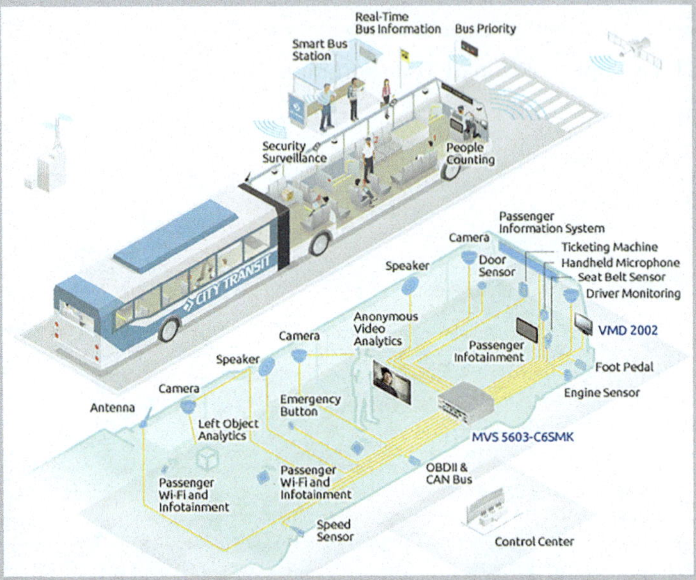

Smart Public Transit

Smart buses provide a solution to the increasing traffic and the demand for streamlined public transportation services. Smart buses offer passengers a convenient and efficient means of traveling, and help bus operators to consolidate fleet management, facilitate daily operations, improve safety and enhance the traveling experience. Equipped with advanced computing, wireless communication, and global navigation satellite system (GNSS), smart buses can be monitored and co-ordinated meticulously to ensure bus services are performing within standards. In addition, real-time live surveillance and video analytics of bus fleets can be implemented to respond to emergency events and ensure security and safety of drivers and passengers. Furthermore, smart buses can monitor and collect data such as driving behavior and passenger flows, giving bus operators insights into its fleet operation and allowing them to make service improvements or timetable rearrangements when necessary.

Available from: https://www.nexcom.com/applications/DetailByDivision/smart-public-transit

Box 1.3 Examples of 'Smart Transportation' System

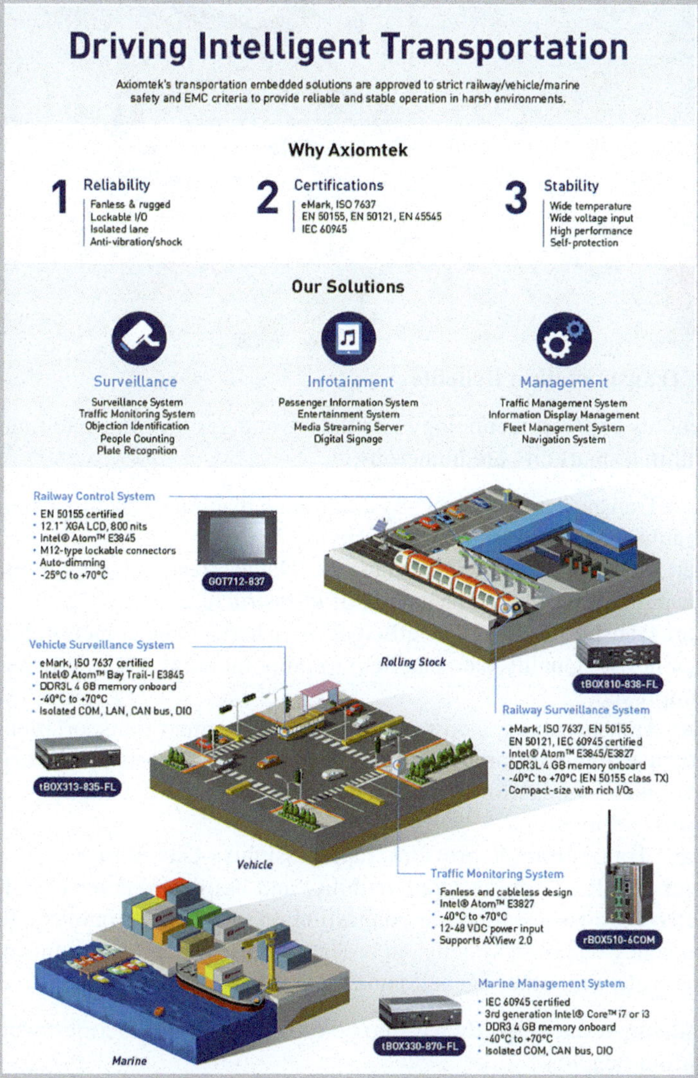

Available from: https://axiomtek.com.tw/eDM/Transportation.html

Box 1.4 Examples of 'Smart Transportation' System

Smart Transportation Benefits

The benefits of smart technology and the advantages they bring to transportation within a smart city are numerous.

- Smart Transportation is safer: By combining machine learning with IoT and 5G, autonomous transportation systems
- Smart Transportation is better managed: Data collection is an important key to responsible public management of infrastructure.
- Smart Transportation is more efficient: With better management comes more efficient use. Quality data can help to pinpoint areas where efficiency can be improved.
- Smart Transportation is cost effective: Because smart transportation makes better use of the resources available, it can cut down costs thanks to preventative maintenance, lower energy consumption, and fewer resources used towards accidents.
- Smart Transportation provides rapid insights: City traffic management centers (TMCs) can get rapid visibility and notifications for trouble spots or city-wide issues affecting congestion on city streets, public safety and emergency response systems, in order to take action or communicate more effectively with other agencies and emergency responders.

Available from: https://www.digi.com/blog/post/introduction-to-smart-transportation-benefits

Box 1.5 Examples of 'Smart Transportation' System

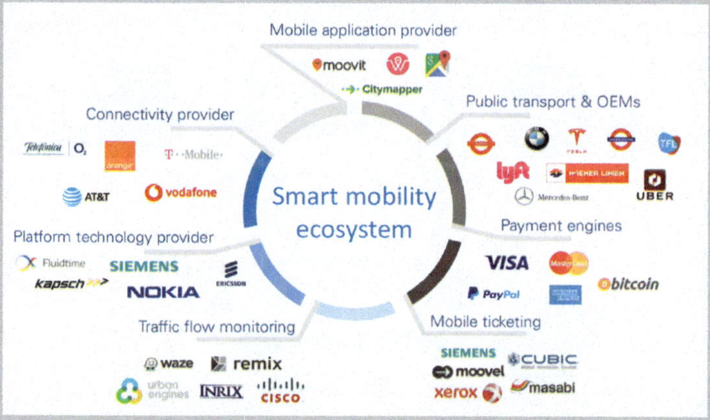

Integrated Mobility Platforms

An Integrated Mobility Platform (IMP) is a key solution to accommodate these customer preferences under a unified digital roof. By integrating different modes of transport, IMPs drastically simplify route planning, making traveling more efficient while providing a customized offer based on selected preferences. As IMPs also feed data back into smart-mobility back-end applications, thereby supporting future infrastructure development, these platforms will increasingly become the nucleus of modern mobility ecosystems.

Smart-mobility ecosystem players, depending on their nature and competitive positioning, have a strategic incentive either to join existing platforms or to establish their own. Whereas private companies are aiming to increase their user bases to grow revenues, public companies generally have stronger interest in promoting mass transit. Similarly, automotive OEMs are seeking to capture a large share of the car-sharing market and push their own vehicles into municipal transportation systems. For large tech companies, such as Google, which seek to expand their map services, the data collection aspect of an IMP is of the most value.

Available from: https://www.adlittle.com/at-en/insights/viewpoints/integrated-mobility-platforms

Box 1.6 Examples of 'Smart Healthcare' System

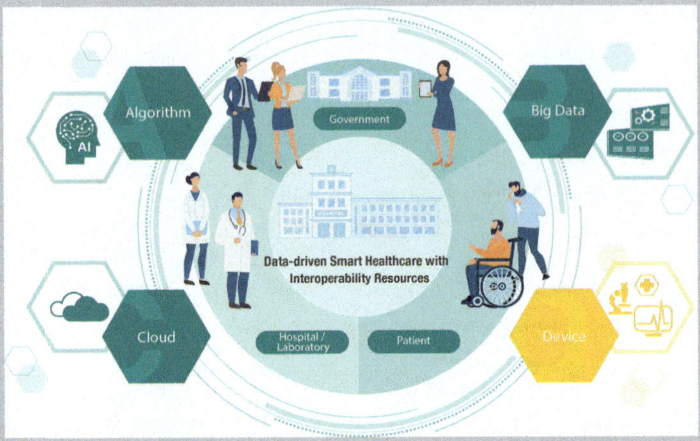

Transform to Data-Driven Smart Healthcare

Prepare for a future of data with continuous data monitoring and collection.

- Automatically collect, integrate, and analyse patient biological data around the clock to establish a research database.
- Instantly deliver data to handheld devices used by doctors and nurses. Even when doctors are not on site, they can immediately understand their patients' conditions, thus reducing treatment response time.
- Doctors can remotely and instantly grasp patients' biological data and case history information.
- No need to manually record data, freeing up time otherwise spent on copying data.
- Medical staff can focus on tasks such as caring for the patients, performing diagnoses, and providing treatment. This reduces patient and family anxiety while enhancing medical service quality.
- Enhance transmission precautions.

Available from: https://madison.tech/data-driven-smart-healthcare/

Box 1.7 Examples of 'Smart Healthcare' System

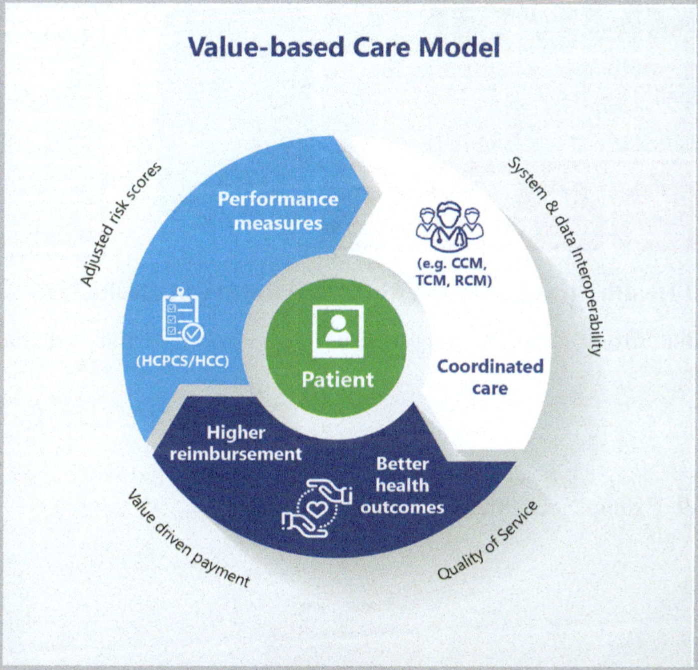

Value-Based Care Model

Telehealth is not the newest healthcare advancement; however, COVID-19 is changing how it is used to treat patients. Telehealth is being utilised during the pandemic to combat the personal protective equipment (PPE) shortage, keep patients isolated and safe in their homes, which allows hospital beds to be open to the patients that really need them.

…Quality measures and quality health care matter and are at the forefront of patient care. Quality measures work to optimise health outcomes by improving quality and transforming the health care system. The measures are able work toward transforming health care because providers and their organisation are monetarily credited for the ability to reduce the amount and probability for return visit.

Available from: https://acehealthcaresolutions.com/value-based-care-and-quality-measures-benefits-providers-and-patients/

Box 1.8 Examples of 'Smart Healthcare' System

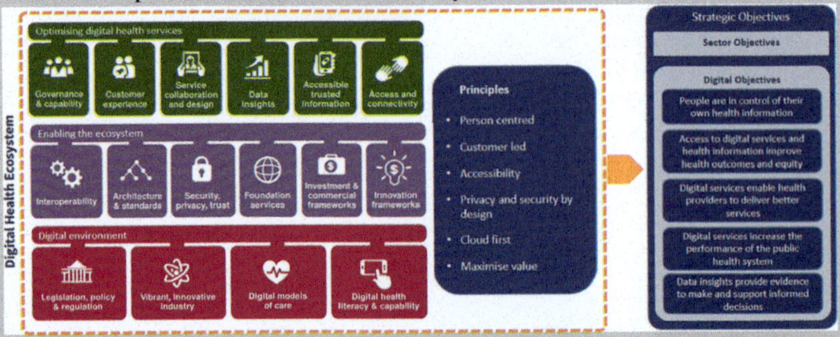

Digital Health Strategic Framework, by Ministry of Health NZ

Available from: https://www.tewhatuora.govt.nz/our-health-system/digital-health

Box 1.9 Examples of 'Smart Healthcare' System

Telehealth

As telehealth continues to expand and improve access to care and increase efficiency, it will become essential to define measures of care. Research is being conducted to identify key performance indicators. Some of the measures will include cross-continuum metrics such as the rate of readmissions, admissions, and ER visits related to telehealth. Additionally, provider, nursing, and patient satisfaction will also provide valuable information as well as financial indicators and health outcomes. The U.S. Department of Health and Human Services called the National Quality Forum to recommend various indicators to measure the use of telehealth as a method of providing care to patients. Healthcare reform is an unending process that nurses have the ability to influence. As telehealth continues to expand, nurses will be phenomenal champions for telehealth. It is vital for nurses to identify and advocate for new opportunities to reach patients in the communities they reside in through telehealth. In doing so, nurses will close the gaps in healthcare delivery and reduce health disparities by stepping forward and utilizing the breadth of their skills to adapt, adopt and implement telehealth resources and services into the mainstream.

Available from: https://app.sophia.org/tutorials/telehealth-scope-and-standards-of-practice-summary.

Box 1.10 Examples of 'Smart Healthcare' System

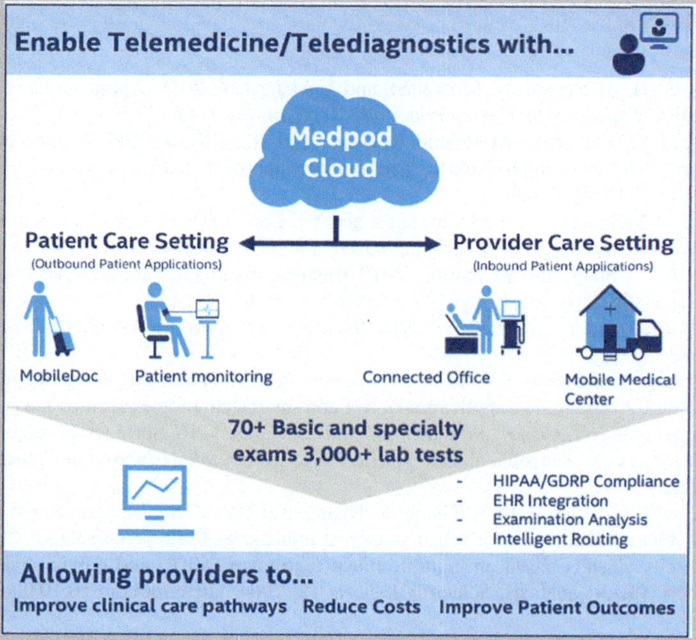

Telehealth: The New Normal for Patient Care

- **MobileDoc**, a telemedicine-enabled diagnostic cart shrunken down to fit into a carry-on-size case that enables professional-level diagnostic capabilities in any environment, allowing providers to accurately diagnose and treat their patients remotely.
- **Provider dashboard with "Single View"**, allowing providers to login once and conduct full telediagnostics consultations using Medpod carts and MobileDocs as well as a Visit feature that enables browser-based virtual consultations on the patient's own device—anytime, anywhere.
- **Remote patient monitoring system** designed to aid chronic disease management and post-acute or inpatient oversight, as part of a value-based healthcare program.
- **Telediagnostics system** that enables remote care on par with a physical visit.

Available from: https://www.insight.tech/5g/telehealth-the-new-normal-for-patient-care#main-content

References

Al Nuaimi, E., H. Al Neyadi, N. Mohamed, and J. Al-Jaroodi. 2015. Applications of big data to smart cities. *Journal of Internet Services and Applications* 6 (1): 1–15.

Al-Ali, A.R., I.A. Zualkernan, M. Rashid, R. Gupta, and M. AliKarar. 2017. A smart home energy management system using IoT and big data analytics approach. *IEEE Transactions on Consumer Electronics* 63 (4): 426–434.

Avazov, K., M. Mukhiddinov, F. Makhmudov, and Y.I. Cho. 2021. Fire detection method in smart city environments using a deep-learning-based approach. *Electronics* 11 (1): 73.

Chen, H., R.H.L. Chiang, and V.C. Storey. 2019. Business intelligence and analytics: From big data to big impact. *MIS Quarterly* 43 (3): 823–840.

Chen, M., S. Mao, and Y. Liu. 2014. Big data: A survey. *Mobile Networks and Applications* 19 (2): 171–209.

Cheshmehzangi, A. 2022a. *ICT, Cities, and Reaching Positive Peace*. Singapore: Springer.

Cheshmehzangi, A. 2022b. The application of ICT and smart technologies in cities and communities: An overview. In *ICT, Cities, and Reaching Positive Peace*, 1–16. Springer, Singapore.

Cheshmehzangi, A., A. Dawodu, and A. Sharifi. 2021. *Sustainable Urbanism in China*. New York: Routledge.

Cheshmehzangi, A., Y. Li, H. Li, S. Zhang, X. Huang, X. Chen, Z. Su, M. Sedrez, and A. Dawodu. 2022. A hierarchical study for urban statistical indicators on the prevalence of COVID-19 in Chinese city clusters based on multiple linear regression (MLR) and polynomial best subset regression (PBSR) analysis. *Scientific Reports* 12: 1964. https://doi.org/10.1038/s41598-022-05859-8.

Cheshmehzangi, A., Z. Su, and T. Zou. 2023. ICT applications and the COVID-19 pandemic: Impacts on the individual's digital data, digital privacy, and data protection. *Frontiers in Human*

Dynamics, Section on Digital Impacts 5. Available from: https://doi.org/10.3389/fhumd.2023.971504

Davenport, T. H. 2014. *Big Data at Work: Dispelling the Myths, Uncovering the Opportunities.* Harvard Business Review Press.

Fan, W., and A. Bifet. 2013. Mining big data: Current status, and forecast to the future. *ACM SIGKDD Explorations Newsletter* 14 (2): 1–5.

Fathi, S., R. Srinivasan, A. Fenner, and S. Fathi. 2020. Machine learning applications in urban building energy performance forecasting: A systematic review. *Renewable and Sustainable Energy Reviews* 133: 110287.

Feng, M., J. Zheng, J. Ren, A. Hussain, X. Li, Y. Xi, and Q. Liu. 2019. Big data analytics and mining for effective visualization and trends forecasting of crime data. *IEEE Access* 7: 106111–106123.

Gandomi, A., and M. Haider. 2015. Beyond the hype: Big data concepts, methods, and analytics. *International Journal of Information Management* 35 (2): 137–144.

Han, J., E. Haihong, G. Le, and J. Du. 2011. Survey on NoSQL database. In *2011 6th International Conference on Pervasive Computing and Applications*, 363–366. IEEE.

Kaisler, S., F. Armour, J.A. Espinosa, and W. Money. 2013. Big data: Issues and challenges moving forward. In *Proceedings of the 46th Hawaii International Conference on System Sciences*, 995–1004.

Kar, P., Z. Xue, S. Pourroostaei Ardakani, and F.C. Kwong. 2022. Are fake images bothering you on social network? Let us detect them using recurrent neural network. *IEEE Transaction on Computational Social Systems* 10 (2): 783–794. https://doi.org/10.1109/TCSS.2022.3159709.

Karkouch, A., H. Mousannif, H. Al Moatassime, and T. Noel. 2016. Data quality in internet of things: A state-of-the-art survey. *Journal of Network and Computer Applications* 73: 57–81.

Khanra, S., A. Dhir, A.N. Islam, and M. Mäntymäki. 2020. Big data analytics in healthcare: A systematic literature review. *Enterprise Information Systems* 14 (7): 878–912.

Kitchin, R. 2014. Big data, new epistemologies and paradigm shifts. *Big Data & Society* 1 (1): 1–12.

Kitchin, R. 2017. Thinking critically about and researching algorithms. *Information, Communication & Society* 20 (1): 14–29.

Kitchin, R., and G. McArdle. 2016. What makes big data, big data? Exploring the ontological characteristics of 26 datasets. *Big Data & Society* 3 (1): 1–10.

Kruse, C.S., R. Goswamy, Y.J. Raval, and S. Marawi. 2016. Challenges and opportunities of big data in health care: A systematic review. *JMIR Medical Informatics* 4 (4): e5359.

Manyika, J., M. Chui, and B. Brown. 2011a. Are you ready for the era of 'big data'? *McKinsey Quarterly* 4: 24–35.

Manyika, J., M. Chui, B. Brown, J. Bughin, R. Dobbs, C. Roxburgh, and A.H. Byers. 2011b. *Big Data: The Next Frontier for Innovation, Competition, and Productivity.* McKinsey Global Institute.

Marr, B. 2015. *Big Data: Using Smart Big Data, Analytics and Metrics to Make Better Decisions and Improve Performance.* John Wiley & Sons.

Mazur, S. 2020. *An introduction to smart transportation: Benefits and examples*, Available from: https://www.digi.com/blog/post/introduction-to-smart-transportation-benefits

Mohler, G.O., M.B. Short, P.J. Brantingham, F.P. Schoenberg, and G.E. Tita. 2011. Self-exciting point process modeling of crime. *Journal of the American Statistical Association* 106 (493): 100–108.

Nan, K., S. Hu, H. Luo, P. Wong, and S. Pourroostaei Ardakani. 2023. A semi-supervised learning application for hand posture classification. In *Big Data Technologies and Applications. BDTA BDTA 2022 2021. Lecture Notes of the Institute for Computer Sciences, Social Informatics and Telecommunications Engineering*, ed. R. Hou, H. Huang, D. Zeng, G. Xia, K.K. Ghany, H.M. Zawbaa, vol. 480. Springer, Cham. https://doi.org/10.1007/978-3-031-33614-0_10

Pham, Q.V., M. Liyanage, N. Deepa, M. VVSS, S. Reddy, P.K.R. Maddikunta, N. Khare, T.R. Gadekallu, and W.J. Hwang. 2021. Deep learning for intelligent demand response and smart grids: A comprehensive survey. arXiv preprint arXiv:2101.08013.

Pourroostaei Adakani, S., N. Du, C. Lin, J. Yang, Z. Bi, and L. Chen. 2023. A federated learning-enabled predictive analysis to forecast stock market trends. *Journal of Ambient Intelligence and Humanized Computing* 14: 4529–453. https://doi.org/10.1007/s12652-023-04570-4

Pourroostaei Ardakani, S., and A. Cheshmehzangi. 2023. *Big Data Analytics for Smart Urban Systems*. Springer, Singapore. In press.

Pourroostaei Ardakani, S., F. C. Knowg, P. Kar, Q. Liu, and L. Li. 2021a. CNN: A cluster-based named data routing for vehicular networks. *IEEE Access, 9,* https://doi.org/10.1109/ACCESS.2021.3131198

Pourroostaei Ardakani, S., C. Zhou, X. Wu, Y. Ma and J. Che. 2021b. A data-driven affective text classification analysis. In *2021 20th IEEE International Conference on Machine Learning and Applications (ICMLA), Pasadena, CA, USA, 13–16 December*. https://doi.org/10.1109/ICMLA52953.2021.00038.

Pourroostaei Ardakani, S. (2017). MSAS: An M-mental health care system for automatic stress detection. *Clinical Psychology Studies* 7 (28): 72–80.

Pramanik, M.I., R.Y. Lau, H. Demirkan, and M.A.K. Azad. 2017. Smart health: Big data enabled health paradigm within smart cities. *Expert Systems with Applications* 87: 370–383.

Provost, F., and T. Fawcett. 2013. Data science and its relationship to big data and data-driven decision making. *Big Data* 1 (1): 51–59.

Shvachko, K., H. Kuang, S. Radia, and R. Chansler. 2010. The Hadoop distributed file system. In *Proceedings of the 2010 IEEE 26th Symposium on Mass Storage Systems and Technologies*, 1–10.

A. M. Townsend. 2013. *Smart Cities: Big Data, Civic Hackers, and the Quest for a New Utopia.* WW Norton & Company.

Ullah, Z., F. Al-Turjman, L. Mostarda, and R. Gagliardi. 2020. Applications of artificial intelligence and machine learning in smart cities. *Computer Communications* 154: 313–323.

World Economic Forum. 2021. *These smart technologies are transforming healthcare.* Available from: https://www.weforum.org/agenda/2021/10/smart-technologies-transforming-healthcare/

Wu, X., X. Zhu, G.Q. Wu, and W. Ding. 2014. Data mining with big data. *IEEE Transactions on Knowledge and Data Engineering* 26 (1): 97–107.

Yamagata, Y., P. P. Yang, S. Chang, M. B. Tobey, R. B. Binder, P. J. Fourie, P. Jittrapirom, T. Kobashi, T. Yoshida, and J. Aleksejeva. 2020. Urban systems and the role of big data. In *Urban Systems Design*, 23–58. Elsevier.

Yang, C., Q. Huang, Z. Li, K. Liu, and F. Hu. 2017. Big Data and cloud computing: Innovation opportunities and challenges. *International Journal of Digital Earth* 10 (1): 13–53.

Zhang, Y., T. Huang, and E.F. Bompard. 2018. Big data analytics in smart grids: A review. *Energy Informatics* 1 (1): 1–24.

Zhou, K., C. Fu, and S. Yang. 2016. Big data driven smart energy management: From big data to big insights. *Renewable and Sustainable Energy Reviews* 56: 215–225.

Zhu, L., F.R. Yu, Y. Wang, B. Ning, and T. Tang. 2018. Big data analytics in intelligent transportation systems: A survey. *IEEE Transactions on Intelligent Transportation Systems* 20 (1): 383–398.

Zikopoulos, P., C. Eaton, D. deRoos, T. Deutsch, and G. Lapis. 2012. *Understanding Big Data: Analytics for Enterprise Class Hadoop and Streaming Data. McGraw-Hill.*

Part I
Smart Transport

Chapter 2
Big Data Analysis for an Optimised Classification for Flight Status: Prediction Analysis Using Machine Learning Classifiers

Abstract Accurate flight delay forecasting is critical for establishing a more efficient airline industry. A smart system in place would help reducing the negative impacts of such delays. Machine learning-enabled Big data solutions have been widely utilized in recent studies to anticipate aircraft delays. They need a data pre-processing to understand and grasp the relevance of each data attribute. The results of data attribute relevance are used to filter out data deemed relevant to aircraft delays and eliminate the data that was redundant and unsuitable for analysis. This chapter trains linear and polynomial regression models to predict the delay time of a flight. The data analysis algorithm runs on a well-known dataset, which comprises flight data from more than ten US airlines from 2009 to 2019. The result indicates that 97.04% of the predicted result has a difference of fewer than 15 min between the actual value.

Keywords Flight delay prediction · Polynomial regression · Linear regression · Data pre-processing · Pyspark

1 Introduction

Smart transportation systems are the backbone of daily operations in cities and regions, where mobility promotes connectedness and economic development. One of the areas of interest in smart transportation systems is the airline industry, where flights in and out of cities/regions are important in everyday operations. Part of optimising the whole transportation network is based on evaluating and minimising flight delays and cancellations. This is important to reduce economic losses at multiple scales and across various sectors. Flight delay is defined as a flight taking off or arriving later than the scheduled time, which occurs in most airlines around the world, costing enormous economic losses for an airline company and bringing huge inconvenience for passengers (Gui et al. 2019). Figure 1 is an instance chronological graph of Southwest Airlines airline that represents a two-week flight delays in January 2009.

The research work in this chapter was supported by our research team members: Zimo Guo, Liushanchuan He, Zhuopu Wang, Zihan Li.

S. Pourroostaei Ardakani and A. Cheshmehzangi, *Big Data Analytics for Smart Transport and Healthcare Systems*, Urban Sustainability,
https://doi.org/10.1007/978-981-99-6620-2_2

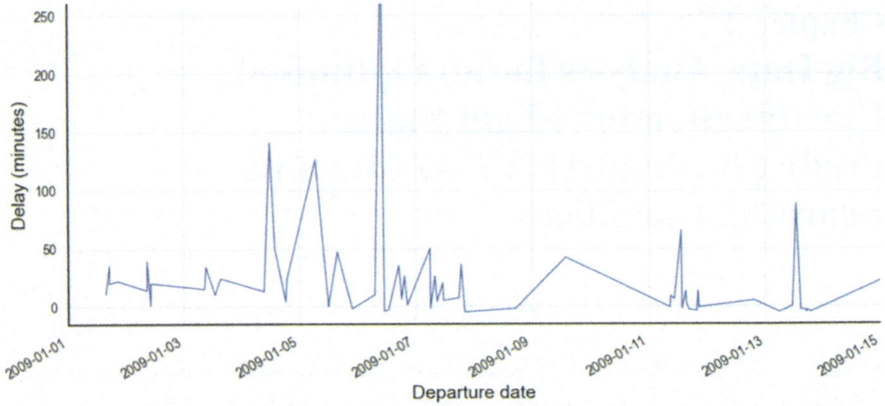

Fig. 1 A two-week sample of Southwest Airlines' flight delay

Studies that have examined the cost of delay show that increasing flight delays end up costing billions of dollars not only to airlines and their passengers but also to society at large (Yimga 2017). This is an economic sustainability issue that should be globally studied. In the United States, flight delays are a severe and common issue. Growing delays endanger the United States' competitiveness in the global economy by limiting the air transportation system's ability to meet the needs of the US economy. The demand for air travel and the increase in the gross domestic product are inextricably linked (Ball et al. 2010). Therefore, the study of flight delays is significantly helpful and meaningful. According to the civil aviation administration of China (CAAC), 47.46% of the delays are caused by severe weather, and 21.14% of the delays are caused by air route problems. The cause of a flight delay is often complex and multidimensional, and various elements need to be analysed, illustrating the relevance between this project and the use of big data.

Big data applications focus on data-driven decision making that form a new shape of technologies in a variety of industries -mainly telecommunication, health-care, and logistics (Huo et al. 2020). Indeed, the overarching area of 'Big Data Analytics' is now an important part of obtaining 'Business Intelligence'. Many firms, particularly large corporations, regard Big data as standard practice. They constantly research the most up-to-date technologies and models to better their Big Data usage. The capacity to show patterns, trends, and relationships, particularly those affecting individuals and enterprises, is one of the most critical expectations from big data analytics to guide meaningful decisions.

This chapter aims to build data regression models to predict/classify flight delays. It uses a huge flight dataset (data 2023) which comprises flight data from more than ten US airlines from 2009 to 2019. There are 28 distinct types of attributes in the data file, each of which records a different type of data. A Big data-enabled approach is used to clean and prepare the dataset, while a linear and a polynomial regression techniques are used to analyse flight data and predict flight delay. The regression results are analysed and discussed to find the best-fitted approach for flight delay regression.

The remainder of this chapter is organised as follows. Section 2 reviews literature on data pre-processing and machine learning applications for flight prediction. Section 3 presents the research principals, while Sect. 4 presents and discusses the experimental results. Section 5 concludes the key findings and addresses future works.

2 Literature Review

Big data applications differ from traditional ones because of data 'Volume', 'Variety', 'Velocity', and 'Veracity' (Chen et al. 2014; Zhang 2015). They have emerged as the leading method for creating, acquiring, processing, and analysing massive volumes of heterogeneous data to capture and recognise data patterns (García et al. 2016). Big data samples are generated during data creation phase. They may include electrophysiology signals, videos and photographs, financial transaction records, stock market indexes, and mobile phone GPS position data (Han et al. 2014).

Pre-processing is the process of evaluating and improving data quality that commonly used in Big data applications. It consists of a series of procedures that turn raw data produced from data extraction into a "clean" and "tidy" dataset. Pre-processing can be recursive, with this set of processes being repeated until the dataset is sufficiently structured for descriptive statistics. Hence, it is important to avoid accidentally introducing bias during pre-processing by altering the dataset in ways that may influence the outcome of research tests (Malley et al. 2016). Pre-processing usually consists of the following (JOSHI and PATEL 2020; Alasadi and Bhaya 2017):

- "Data cleaning"—It removes incomplete information, noise, outliers, and redundant or inaccurate records from the database while minimising bias introduction.
- "Data integration"—It consolidates the multiple raw databases into a unique dataset that contains all of the data needed for the statistical analysis.
- "Data transformation"—Transforms or scales variables contained in a variety of formats or units in the raw data into forms or measures that are more relevant to the quantitative research methodology.
- "Data reduction"—This reduces unnecessary information and variables, as well as reorganises the data in an efficient and "tidy" way for analysis, once the database has been incorporated and converted.

Data visualisation is the process to represent data behaviours, and depict and model data correlation. It helps data scientists have a better understanding of the data overview, which is usually essential for optimisation and decision making. According to Evergreen (2019), data are more persuasive when modelled and depicted via graphs. Therefore, pursuing high-quality graphics to visualise the data is of vital importance. For example, heat-map visualisation is a technique that is frequently used to visualise data from microarrays, which is a biological approach for determining the degrees of gene activation in large groups of cells (Pryke et al. 2007). In this application, a typical dataset has a few dozen samples and hundreds, if not thousands, of genes.

Big data visualisation differs from traditional ones as the dataset usually has large volume, high velocity, high variety, and low veracity. For instance, traditional visualisation methods are becoming obsolete in Big data applications due to very fast data accumulation. For this, Big data visualisation techniques aim to find intriguing patterns and relationships in large amounts of data with careful and precise feature section and dimension reduction (Ali et al. 2016).

2.1 Machine Learning for Flight Predictions

Flight delay plays a key role in flight passenger dissatisfaction, and airlines' financial losses. It is a very common global issue during high seasons and/or in big airports. As (Sternberg et al. 2017) reports, 36% of flights in Europe were delayed by more than five minutes, 31.1% of flights in the United States were delayed by more than 15 min, and 16.3% of flights in Brazil were canceled or experienced delays of more than 30 min in 2013. Hence, this needs a careful monitoring and analysis to minimise delays and their impacts on flights and passengers. Machine learning-enabled flight delay prediction has the capacity to minimise the impact of flight delays on passenger, airlines, airports and flights. According to the airline statistics (Jiang et al. 2020), flight delays in 79% of 2019 air-travels result in tens of billions dollars direct and indirect financial losses.

There are several machine learning approaches for flight delay prediction that have been introduced in recent years. (Gui et al. 2019) introduces an Long Short-Term Memory (LSTM) machine learning method that runs on Automatic Dependent Surveillance-Broadcast (ADS-B) Message-Based Aviation Big Data Platform. The ADS-B system is an integrated communication and surveillance system for air traffic management (ATM), in which flights broadcast their location and other information on the same frequency band regularly. This research divides the input into several attributes, including flight information, flight time, weather, airport, and air route. Then, it classifies the delay into four classes, no delay, within one hour, within two hours, and over two hours, and train an LSTM model that reaches a prediction accuracy of 85%. Kim et al. (2016) implements an Recurrent Neural Network (RNN) method that runs on day-to-day sequence data on a specific airport to predict the delay of that airport. This approach divides the output into two classes according to the delay threshold (15 and 30 min). The neural network takes a sequence of flight information as the input and trains the RNN model. This approach gives a maximum accuracy of 87.42%. Regression is also popular for flight delay prediction. Yang et al. (2018) propose an Support Vector Machine (SVM) regressions to predict flight delay. This is trained using historical data from LAX (Los Angeles International Airport), in the US, and PVG (Pudong International Airport), in Shanghai. The performance of the proposed model is evaluated against a multivariate linear regression. The results show that the SVM regression model gives a better prediction accuracy. Ding (2017) uses multiple linear regression to predict flight delay in 2017. It selects similar records based on departure airport, flight type, and weather type, divides the records according to the departure time, and uses the data partitions to train the predictive model. The performance of the linear regression model is measured and compared with Naive Bayes and Decision Tree models. The results show that the proposed linear regression model gives a better flight delay prediction as compared to the benchmarks. Choi et al. (2016) evaluate the performance of K-Nearest Neighbor (KNN), Random Forest regression, Decision Tree, and LTSM in flight delay predictions. The results show that Random Forest gives the best prediction as compared to KNN, Decision Tree, and LTSM (Gui et al. 2019). Rebollo and Balakrishnan (2014) aims to improve the Random Forest model. It also introduces spatial variables, such as the past flight delay of the departure place, and the recent flight delay of the airline, in addition to time variables and uses them in model training. Pamplona et al. (2018) proposes an Artificial Neural Network (ANN) model to predict flight delays. It uses Random Search technique to find the best-fitted hyper-parameters and optimise the prediction performance. According to the results, this approach achieves a prediction accuracy of 90% in flight delay prediction. Guo et al. (2021) proposes a hybrid method of Random Forest Regression and Maximal Information Coefficient (RFR-MIC) for flight departure delay prediction. The results show a better results as compared to KNN, Random forest, Linear regression, and ANN. Yu et al. (2019) proposes a prediction model as the combination of DNB (novel deep belief network) and SVR (Support vector regression). DBN is used to extract key factors and eliminate redundant information, while SVR model processes the variables with indirect relations (which can not be handled in a linear prediction). The results show a strong model robustness based on different flight departures, and a much better accuracy as compared to KNN and linear regression.

3 Methodology

This section proposes a data pre-processing approach to study the data correlations, clean the records, extract the features, and prepare the dataset for machine learning. Data pre-processing is required in this study as this is a Big data analysis and the original dataset is huge. Knowledge discovery and data pattern recognition becomes complicated if the dataset contains a large amount of duplicate, unrelated, noisy and untrustworthy data elements (Alexandropoulos et al. 2019). This issue is popular in Big data analysis and therefore Big datasets need pre-processing to be ready for further analysis (i.e., machine learning).

Apache Spark (pyspark) data processing framework is used to build the data pre-processing approach. By this, unnecessary features are eliminated (e.g., df.drop), missing values are removed (e.g., df.dropna), total number of flights of each airline

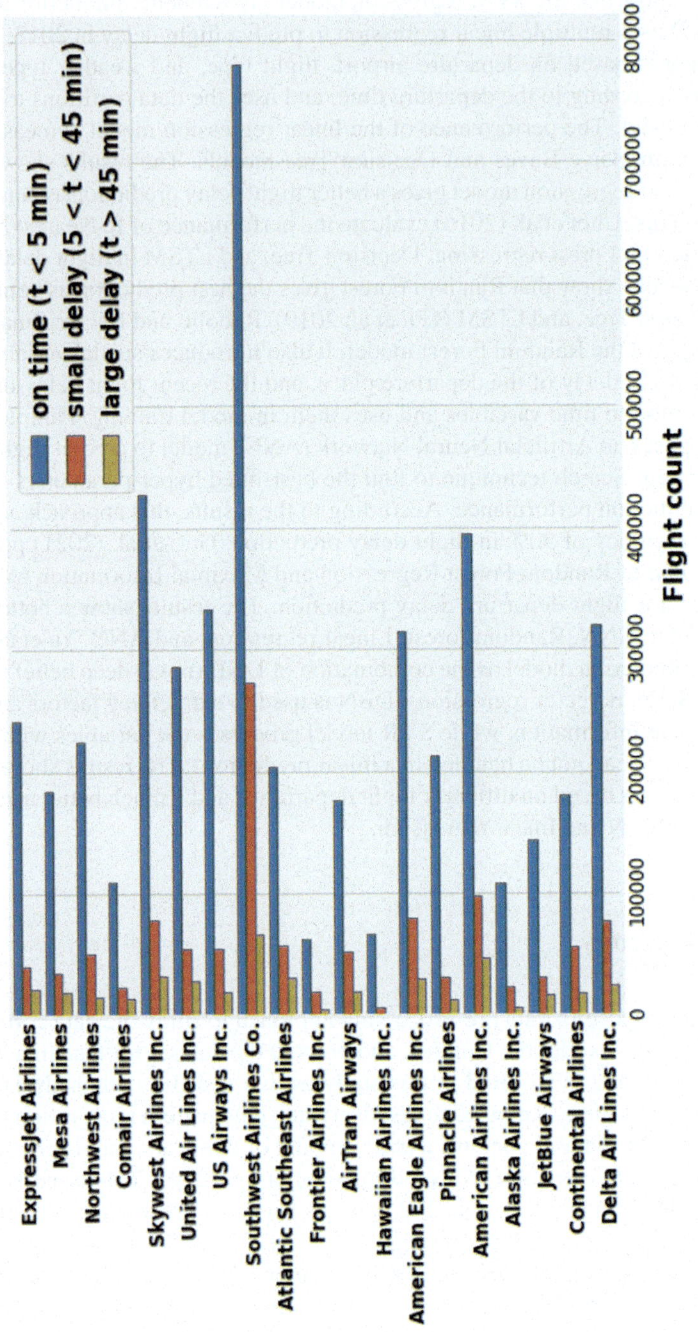

Fig. 2 Flight delay classes

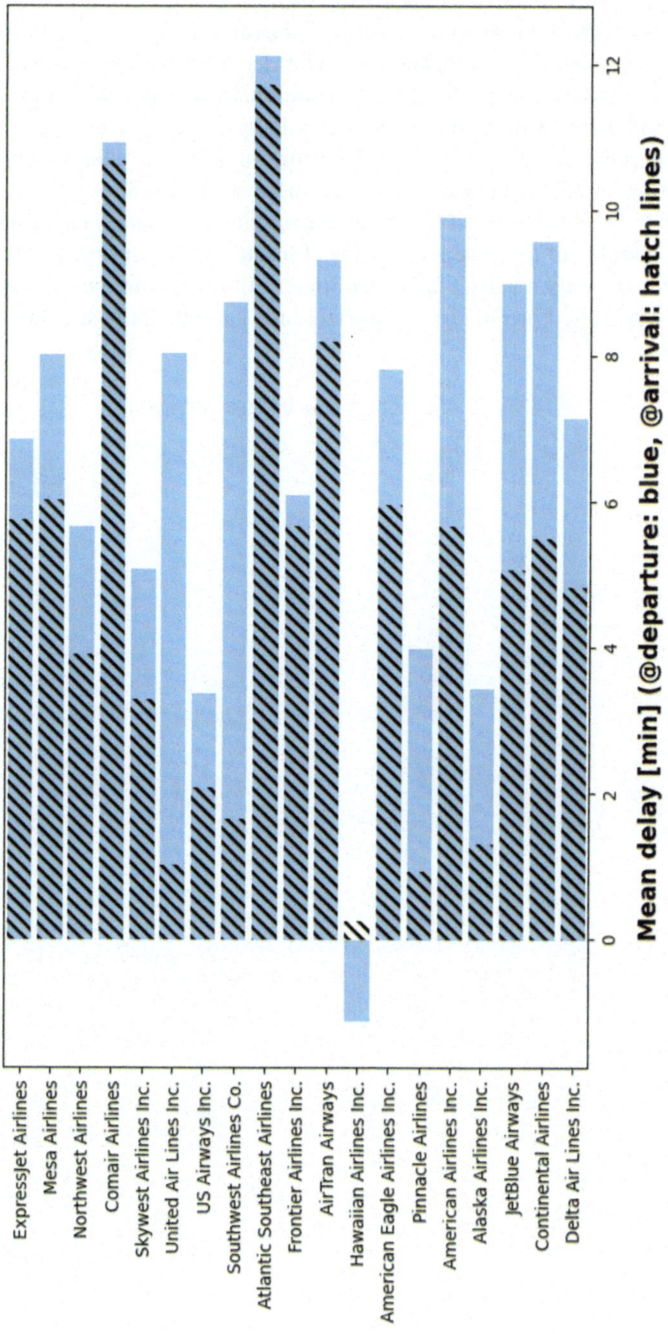

Fig. 3 Flight delays: arrival, departure, and mean

is calculated and sorted, maximum/minimum/average values of departure delays are measured, and partitioned data frames are joined.

Figure 2 shows the delays less than 5 min, in the range of 5 <t <45 min, and greater than 45 min. This delay is calculated based on departure delay, however, it differs from what recorded on arrival. By this, the differences between departure (predicted by airlines) and arrival (the actual) delays should be calculated and studied.

Figure 3 depicts the differences of departure and arrival delays, where arrival delays are often less than departure delays. It supports that airlines alter flight speeds in an attempt to lessen arrival delays. Nonetheless, departure delays at different times of the delay could lead to subsequent delays for later flights during the day.

Figure 4 shows only a subset heat map (due to the large number of airports) that gives the most delayed origin airport (horizontally) and the largest delay in specific

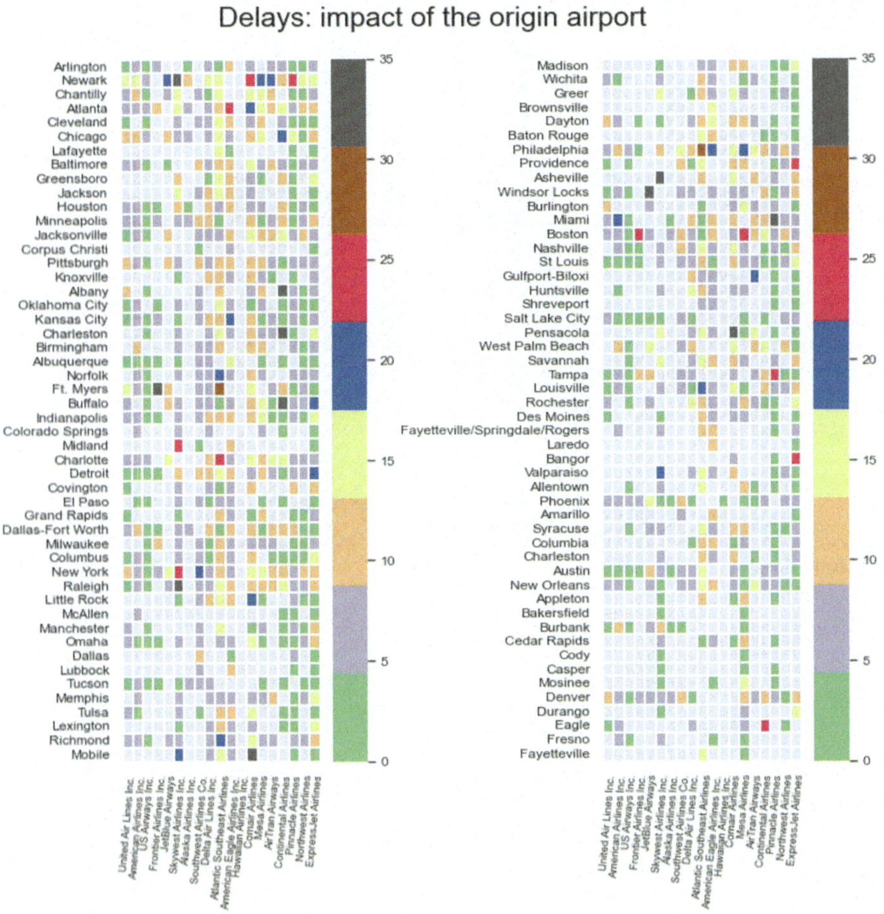

Fig. 4 Flight delays: the impact of the origin airport

airlines (vertically). It helps to recognise flight delay patterns if the data is modelled and studied chronologically.

Linear and polynomial regression methods are used to predict flight delays. The models are trained using the prepared dataset, and then tested and compared to evaluate the prediction results. A 5-fold cross-validation approach is also used to avoid model over-fitting.

4 Result and Discussion

This section provides the experimental results and discusses the research findings. Figure 5 shows the total flight number for a sample set of airlines. It calculates the maximum, minimum, and average values of departure delays and the maximum, minimum, and average values of arrival delays for each airline.

Figures 6 and 7 show the flight information for each company. Indeed, they include seven pie charts to illustrate the detailed relationships between departure delays and airline companies and the relationship between arrival delays and airline companies.

The difference between airlines and departure places has a great impact on flight delay. Hence, the predictive model is built using the flight data of the same airline and departure place to minimise the impact of the airline and departure places.

Data pre-processing results in an increased data quality. Figure 8 shows the differences of the calculated average delay of Southern Airlines (WN) flight from Detroit Metropolitan Wayne County Airport (DTW) between the original and prepared datasets. The blue line represents the average delay calculated from the original data, and the green line represents the average delay after data pre-processing. This supports that a better result is achieved if the dataset is preprocessed. Indeed, it shows that the average delay of the original data is significantly higher than the preprocessed statistics at any time in a day.

Figure 9 shows the machine learning results and the fitting degree of the regression curve when the polynomial order is 1, 2, and 3. Indeed, it compares the impact of the different degrees of a polynomial on model performance. According to the results, high polynomial order leads to over-fitting, while low order results to the lack of model fitness.

OP_CARRIER	count	max(DEP_DELAY)	avg(DEP_DELAY)	min(DEP_DELAY)	max(ARR_DELAY)	avg(ARR_DELAY)	min(ARR_DELAY)
HA	72538	999.0	-1.098775814056081	-50.0	1026.0	0.2607874493368993	-53.0
F9	86425	747.0	6.109239224761354	-30.0	745.0	5.685623372866647	-48.0
AS	135533	814.0	3.450886499966798	-61.0	793.0	1.3170298008603072	-83.0
OH	145470	685.0	10.941802433491441	-50.0	680.0	10.687756925826632	-77.0
B6	191043	814.0	9.016446559151605	-52.0	823.0	5.07928058081165	-92.0

Fig. 5 Airlines' summarised statistics

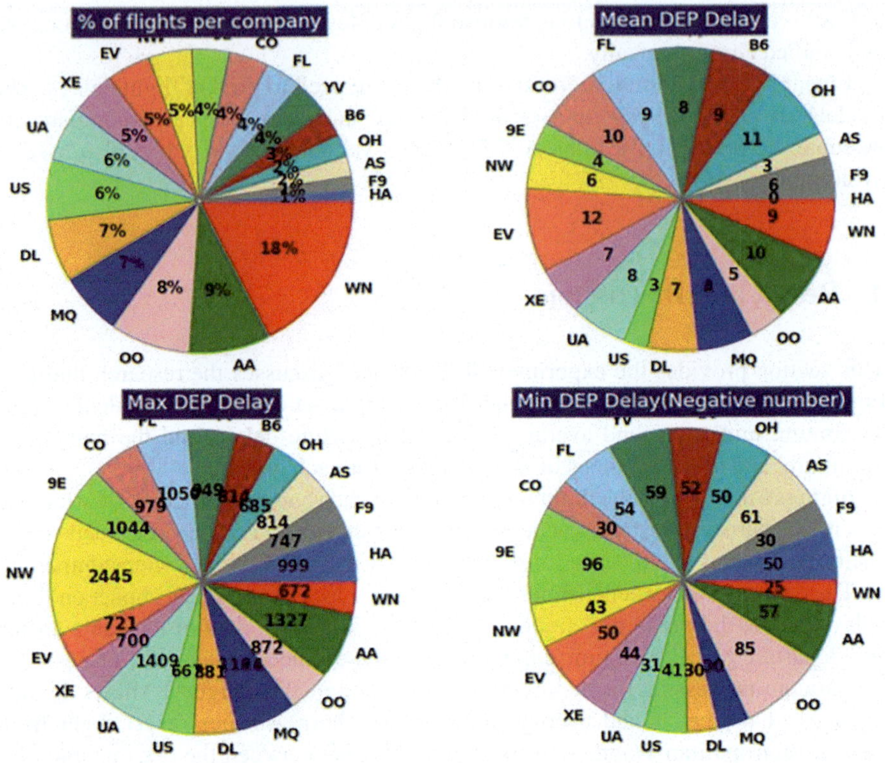

Fig. 6 Departure delay statistics

The result of the linear regression model is shown in Fig. 10. It is generated using the preprocessed flight dataset (i.e., WN Airlines) with the same airline and various departure places as input. The abscissa represents the actual delay, the ordinate is the predicted delay, the green line represents the overall linear trend of the predicted result, and the red line shows the standard curve when the predicted result is consistent with the actual value. According to the results, the variance between the actual and predicted values is 42.65, and 3.15% of the total samples have a difference of more than 15 min delay between the predicted and the actual value.

Figure 11 depicts the results of polynomial regression. The model is trained with a polynomial a polynomial degree of 2. As the results shows, the variance between the actual and predicted values is 41.00, and 2.96% of the total samples have a difference of more than 15 min delay between the predicted and the actual value.

Figure 12 shows the prediction results of the polynomial regression model. According to it, the variance between the actual and predicted values is 80.36, and 7.534% of the total samples have a difference of more than 15 min delay between the predicted and the actual value.

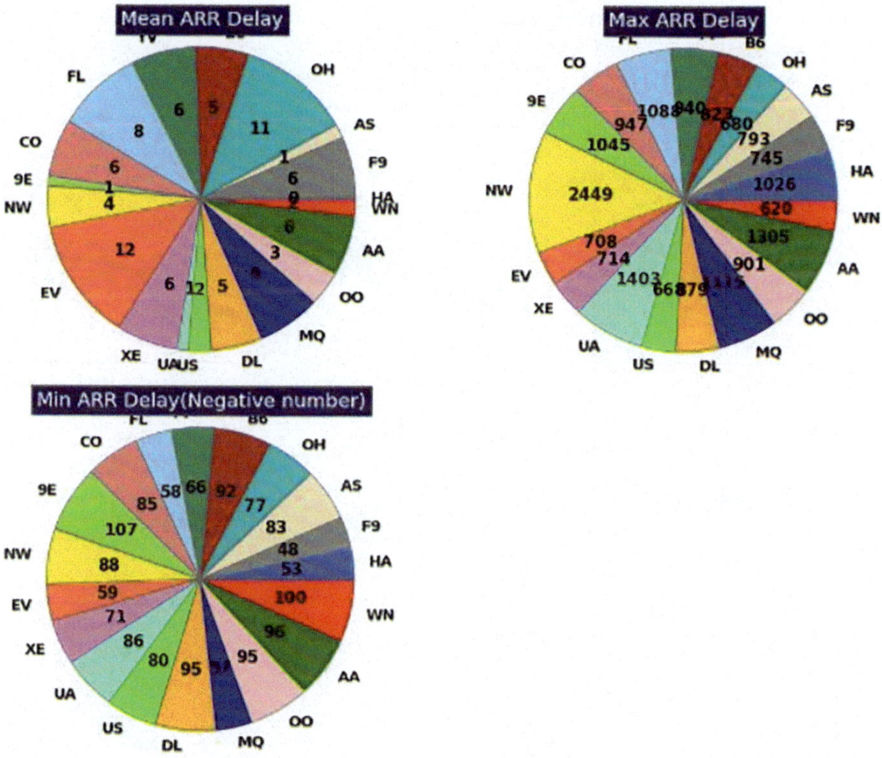

Fig. 7 Arrival delay statistics

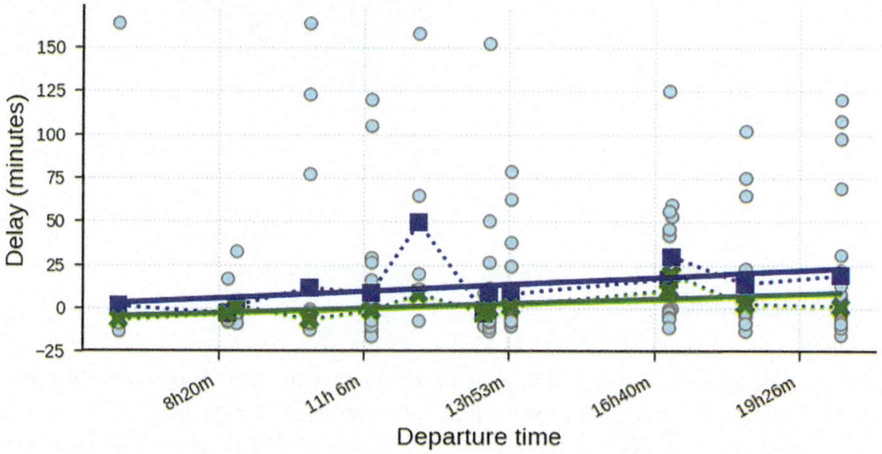

Fig. 8 The differences of the average delay between original and pre-processed datasets

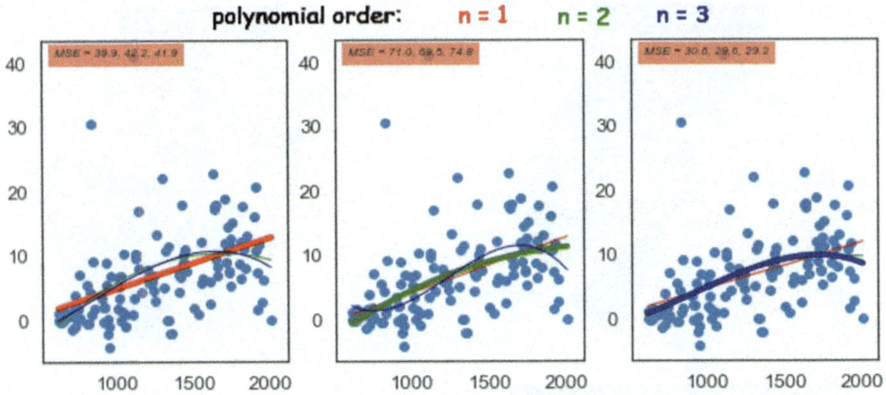

Fig. 9 Average flight delay using the preprocessed dataset

Fig. 10 The result of linear regression

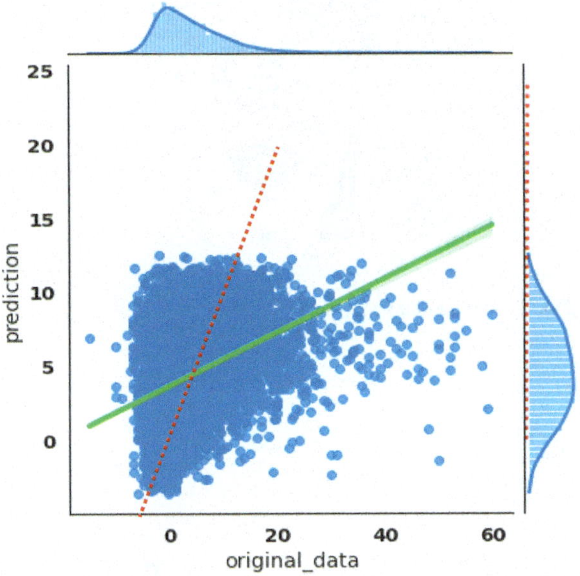

5 Conclusion

This chapter aims to study the influence of different airlines on flight using Big data-enabled machine learning approach. For this, a huge public dataset containing the flight information of 28 airlines for 10 years (2009–2019) is used. This dataset is prepared using a Big data data preprocessing approach on pyspark to clean-up noisy and missing values, extract the key features, and remove the irrelevant records. The prepared dataset is used by linear and polynomial regression method to predict the

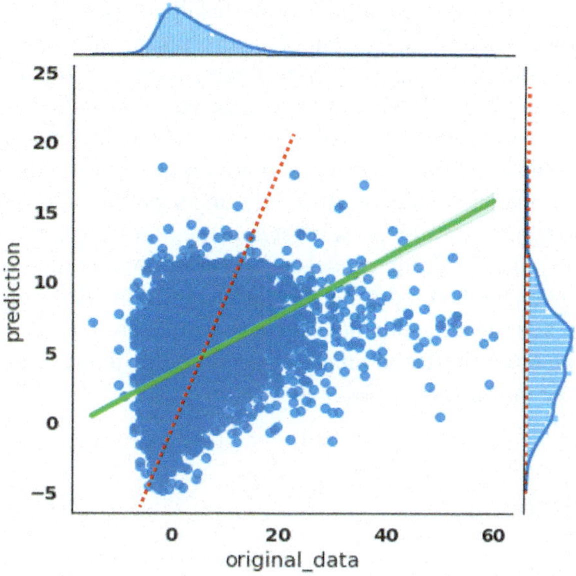

Fig. 11 The result of polynomial regression

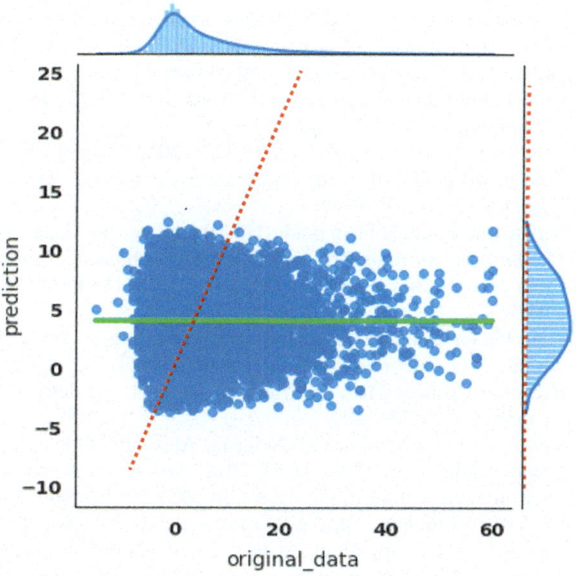

Fig. 12 Test prediction using polynomial regression

flight delays. According to the results of both the models, 92.5% of the predicted results show a delay 15 min against the actual values.

The performance of linear regression is compared to polynomial regression to find the best-fitted predictive approach. According to the results, the overall trend of the prediction results of both models is relatively similar. However, the prediction result of polynomial regression is more spread out, which means the model tends to predict some extreme cases under polynomial regression. The model performs slightly worse on the test set than the training set, but still within a reasonable range. In other words, the model tends is more conservative, which means that the prediction given by the model may be significantly smaller than the actual result in the face of a significant actual delay. These results are mainly because the input data tends to be more similar and more continuous due to the removal of extreme values, making the model less sensitive to extreme cases.

References

Airline delay analysis. https://www.kaggle.com/datasets/sherrytp/airline-delay-analysis. Retrieved January 2023.

Alice Sternberg, Jorge Soares, Diego Carvalho, and Eduardo Ogasawara. 2017. A review on flight delay prediction. arXiv preprint arXiv:1703.06118.

Andy Pryke, Sanaz Mostaghim, and Alireza Nazemi. 2007. Heatmap visualization of population based multi objective algorithms. In *International Conference on Evolutionary Multi-Criterion Optimization*, 361–375. Springer.

ASHISH P. JOSHI and BIRAJ V. PATEL. 2020. Data preprocessing: the techniques for preparing clean and quality data for data analytics process. *Oriental Journal of Computer Science and Technology* 13 (2): 78–81.

Bin Yu, Zhen Guo, Sobhan Asian, Huaizhu Wang, and Gang Chen. 2019. Flight delay prediction for commercial air transport: A deep learning approach. *Transportation Research Part E: Logistics and Transportation Review* 125: 203–221. ISSN 1366-5545. https://doi.org/10.1016/j.tre.2019.03.013.

Brian Malley, Daniele Ramazzotti, and Joy Tzung-yu Wu. 2016. Data pre-processing. *Secondary Analysis of Electronic Health Records* 115–141.

Chen, Min, Shiwen Mao, and Yunhao Liu. 2014. Big data: A survey. *Mobile Networks and Applications* 19 (2): 171–209.

Daniel Alberto Pamplona, Li Weigang, Alexandre Gomes de Barros, Elcio Hideiti Shiguemori, and Claudio Jorge Pinto Alves. 2018. Supervised neural network with multilevel input layers for predicting of air traffic delays. In *2018 International Joint Conference on Neural Networks (IJCNN)*, 1–6. https://doi.org/10.1109/IJCNN.2018.8489511.

Ding, Yi. 2017. Predicting flight delay based on multiple linear regression. *IOP Conference Series: Earth and Environmental Science* 81: 012198. https://doi.org/10.1088/1755-1315/81/1/012198.

Gui, Guan, Fan Liu, Jinlong Sun, Jie Yang, Ziqi Zhou, and Dongxu Zhao. 2019. Flight delay prediction based on aviation big data and machine learning. *IEEE Transactions on Vehicular Technology* 69 (1): 140–150.

Han, Hu., Yonggang Wen, Tat-Seng. Chua, and Xuelong Li. 2014. Toward scalable systems for big data analytics: A technology tutorial. *IEEE access* 2: 652–687.

Jiage Huo, Kin Lok Keung, Carman Ka Man Lee, Kam KH Ng, and KC Li. 2020. The prediction of flight delay: Big data-driven machine learning approach. In *2020 IEEE International Conference on Industrial Engineering and Engineering Management (IEEM)*, 190–194. IEEE.

Juan Jose Rebollo and Hamsa Balakrishnan. 2014. Characterization and prediction of air traffic delays. *Transportation Research Part C: Emerging Technologies* 44: 231–241. ISSN 0968-090X. https://doi.org/10.1016/j.trc.2014.04.007.

Michael Ball, Cynthia Barnhart, Martin Dresner, Mark Hansen, Kevin Neels, AR Odoni, Everett Peterson, Lance Sherry, Antonio Trani, and Bo Zou. 2010. Total delay impact study: A comprehensive assessment of the costs and impacts of flight delay in the United States.

Salvador García, Sergio Ramírez-Gallego, Julián Luengo, José Manuel Benítez, and Francisco Herrera. 2016. Big data preprocessing: methods and prospects. *Big Data Analytics* 1 (1): 1–22.

Stamatios-Aggelos N Alexandropoulos, Sotiris B Kotsiantis, and Michael N Vrahatis. 2019. Data preprocessing in predictive data mining. *The Knowledge Engineering Review* 34.

Stephanie D. H. Evergreen. 2019. *Effective Data Visualization: The Right Chart for the Right Data.* SAGE Publications.

Suad A Alasadi and Wesam S Bhaya. 2017. Review of data preprocessing techniques in data mining. *Journal of Engineering and Applied Sciences* 12 (16): 4102–4107.

Sun Choi, Young Jin Kim, Simon Briceno, and Dimitri Mavris. 2016. Prediction of weather-induced airline delays based on machine learning algorithms. (Online; Accessed 01 May 2022).

Syed Mohd Ali, Noopur Gupta, Gopal Krishna Nayak, and Rakesh Kumar Lenka. 2016. Big data visualization: Tools and challenges. In *2016 2nd International Conference on Contemporary Computing and Informatics (IC3I)*, 656–660. IEEE.

Yang, H.E., Z.H.U. Jinfu, and Z.H.O.U. Qinyan. 2018. Airport flight delay prediction based on svm regression. *Journal of Civil Aviation University of China* 36 (1): 30.

Yimga, Jules. 2017. Airline on-time performance and its effects on consumer choice behavior. *Research in Transportation Economics* 66: 12–25.

Young Jin Kim, Sun Choi, Simon Briceno, and Dimitri Mavris. 2016. A deep learning approach to flight delay prediction. In *2016 IEEE/AIAA 35th Digital Avionics Systems Conference (DASC)*, 1–6. IEEE.

Yushan Jiang, Yongxin Liu, Dahai Liu, and Houbing Song. 2020. Applying machine learning to aviation big data for flight delay prediction. In *2020 IEEE Intl Conf on Dependable, Autonomic and Secure Computing, Intl Conf on Pervasive Intelligence and Computing, Intl Conf on Cloud and Big Data Computing, Intl Conf on Cyber Science and Technology Congress (DASC/PiCom/CBDCom/CyberSciTech)*, 665–672. IEEE.

Zhang, Ke. 2015. Four changes of modern universities from the perspective of "4v" of big data. *Canadian Social Science* 11 (3): 292–297.

Zhen Guo, Bin Yu, Mengyan Hao, Wensi Wang, Yu Jiang, and Fang Zong. 2021. A novel hybrid method for flight departure delay prediction using random forest regression and maximal information coefficient. *Aerospace Science and Technology*, 116: 106822. ISSN 1270-9638. https://doi.org/10.1016/j.ast.2021.106822.

Chapter 3
On-Board Unit Freight Transport Data Analysis and Prediction: Big Data Analysis for Data Pre-processing and Result Accuracy

Abstract Traffic prediction is a complex urban problem that needs to collect and analyse big data and train accurate machine learning models. Solving it will contribute to society's efficient production and development, particularly in optimising urban systems. This chapter aims to build a precise model to predict the number of freight vehicles in future timestamps based on historical data provided On-Board Unit (OBU) datasets in Belgium's road networks. We contributed to two novel solutions to solve the time series prediction task. The first contribution is preprocessing the data using SparkSQL and generating nine features from the prediction timestamps. The second is LSTM and LSTM + FCN deep learning models with tuned parameters and training with the newly generated features. The results of our LSTM model are more accurate than those of other models on this dataset that we have already examined, reaching an accuracy of 99.89%. The preprocessing stage is proven to be vital for the performance of the LSTM model. As a result, models using our preprocessed nine features are much more accurate on this freight transportation prediction task. Discussions are also raised to understand better freight transport prediction tasks from the data and results points of view.

Keywords On-Board Unit · Time series traffic prediction · Freight transport · Urban system · LSTM

1 Introduction

The real-time situation of a road network could be affected by many factors, including road conditions, number of junctions and turns, traffic flow, weather, and types of vehicles—cars, motorcycles, buses, or heavy-good vehicles (Sadeghi-Niaraki et al. 2011). Governments need to monitor, analyse and minimise the adverse effects of heavy-goods vehicles because they are often the primary source of road congestion due to their slow speed, loading time, and big size (Ballantyne et al. 2013).

The research work in this chapter was supported by our research team members: Gabrielle Saputra HADIAN, Mingyan WANG, Yiming TANG, and Yuxin SHI.

S. Pourroostaei Ardakani and A. Cheshmehzangi, *Big Data Analytics for Smart Transport and Healthcare Systems*, Urban Sustainability,
https://doi.org/10.1007/978-981-99-6620-2_3

Among many countries, Belgium has taken the road freight transport issue seriously. As of 2016, owners of Belgian freight vehicles must install an On-Board Unit (OBU), which communicates with the Road Side Units (RSUs) through microwaves at the toll station to record and submit their travel data. The data include timestamps, street identifiers, traffic counts (number of trucks), and traffic speed. We utilised the time series datasets collected by Brussels Mobility, containing Belgium country road network data in three different periods: January–March 2019, March–June 2021, and June–October 2021. Different datasets also exist for different temporal resolutions, i.e., how often OBUs transmit data: 5, 15, 30, and 60 min.

Brussels Mobility collected the data for projects called icity.brussels and MOBIlity-Advanced Indicators Dashboard (MOBI-AID). MOBI-AID project (ULB 2023) is a monitoring system project aiming to help the government obtain better insights into the dynamic mobility in Brussels and plan suitable policies for the region. MOBI-AID project has reported various challenges related to their data, ranging from data storage, data access, and the size of the big data. In order to tackle these problems, we had two major contributions to this chapter. Our first contribution is the novel solution to solve the big data problem—preprocessing the data using SparkSQL, and analysing different factors (features) that affect the number of trucks at a specific time in Belgium. The second contribution is to create and improve an LSTM deep learning model with our highly tuned parameters and a processed dataset to forecast the number of trucks in a future timestamp based on historical data and patterns. Such approach is important for further analysis and optimisation processes of urban systems, such as transportation networks or systems.

In response to the above matters, we first cleaned the raw data by removing all the missing values and duplicates. Then, we performed feature engineering by converting the timestamps into meaningful features for the prediction. In this chapter, a long-term short memory network (LSTM) is used to model the freight transportation of Belgium. The LSTM is suitable for sequential data and has fast training time compared with other machine learning models (Li et al. 2021). Our model has reached a 0.9989 R-Squared value, proving that this LSTM model fits the observed data. Therefore, the LSTM can achieve outstanding accuracy for traffic prediction tasks and is simpler and easier to train than other complex models.

This study's general outline is as follows: the following section presents the literature review, containing existing research and literature on the same topic. Section 3 defines the details and reasoning behind our implementation. Section 4 contains our results, i.e., our interpretation, explanations, and visualisations. Lastly, Sect. 5 is the conclusion, summarising the project's findings and ideas for future work.

2 Literature Review

Time series analysis and forecasting are widely used in different areas, including road network traffic prediction, and could be further divided into univariate and multivariate analyses. The difference between univariate and multivariate time series is based on the number of variables used in the observation (Beeram and Kuchibhotla 2020).

In a univariate time series classification task, the input is a vector X containing k observations, and the output is a value y. In multivariate time series classification (MTSC), each element in vector X will have d dimensions. This is where the word 'multivariate' comes from. Multivariate time series data could be defined as sequences of data points from multiple variables collected every regular interval in a time domain (Karim et al. 2019). This chapter's multivariate time series classification (MTSC) aims to analyse the dataset, find out the relations between different variables (specifically between the date-time components and the traffic count), and finally perform the freight traffic count prediction. There are various methods to solve MTSC tasks, including distance measures, shapelets, interval summarising, and machine learning.

Dynamic time warping (DTW) is a popular distance-based approach for MTSC problems, usually taken as a baseline model for comparison. Shokoohi-Yekta et al. (2017) proposed approaches to adapt DTW to solve MTSC tasks. Similar to dynamic time warping, which adapts univariate classification problems to multivariate problems, ensembles of univariate classifiers are common. They are heavily used and are the most popular benchmark in TSC.

Bagnall et al. (2020) built the latest version of HIVE-COTE, which builds models on each input dimension and then puts together all the output of each dimension to produce a single result for MTSC. HIVE-COTE and the later mentioned InceptionTime and ROCKET do not significantly improve accuracy. Considering the speed of training, ROCKET is the fastest, but it performs poorly on specific datasets. However, HIVE-COTE makes the model as controllable as possible and has a close training time to ROCKET. Shapelets are focused on finding a particular local pattern in the sequence data that can occur anywhere in the sequence. A shapelet-based method for MTSC was proposed by Karlsson et al. (2016). This approach applies randomly chosen shapelets into a random forest, increasing the scalability of the shapelets and maintaining interpretability. The model has a long training time, but the prediction accuracy is better than the LTS and cDTW methods, which are a variation of DTW.

Schäfer and Leser (2017) presented a novel MTSC classifier, MEASEL + MUSE, which involves the interplay of different dimensions features and counters the noise and irrelevant data when dealing with high-dimensional data. Middlehurst et al. (2020) built a novel classifier named the 'Canonical Interval Forest (CIF)' by combining the time series tree model classifiers (Deng et al. 2013) and Catch22 (Lubba et al. 2019) features. This classifier improves the accuracy and extends the original separate classifier to the multivariate classification problem. Combining HIVE-COTE and CIF shows significant improvement in the UCR archive.

Dempster et al. (2020) implemented The Random Convolutional Kernel Transform, ROCKET, which applies random convolutional kernels to train a linear classifier. The model could achieve faster and more accurate performance compared with DTW. Le Nguyen et al. (2019) came up with a symbolic representations-based model trying to address variable-length time series. This learner is an extension of the SEQL learner implemented by Le Nguyen et al. (2017). There are three key ideas: (1) obtaining various parameters through symbolic representation, (2) including several domain representations, such as time and frequency domains, and (3) utilising the greedy feature selection approach to navigate in a symbolic-word space efficiently. This method is quite accurate and fast, and requires less memory, but it may not be suitable for some datasets.

As deep learning technology is getting more popular, there are increasing numbers of deep learning approaches to solve time series classification problems. Ismail Fawaz et al. (2019) did thorough research on nearly 9000 deep-learning models on 12 MTSC datasets. Among all the deep learning models in this chapter, the ResNet model proposed by He et al. (2016), which was initially applied to the baseline deep neural network TSC model by Wang et al. (2017), achieved the best performance compared with the Fully Convolutional Network (FCN) and other architectures.

Ismail Fawaz et al. (2020) also introduced their approach for univariate time series problems called InceptionTime, which is an ensemble of deep CNN models inspired by the Inception-v4 architecture (Szegedy et al. 2016). This approach could reach as high accuracy as the HIVE-COTE algorithm while being more scalable and fast to train—making this model achieve the new state-of-the-art performance.

The above two approaches, ResNet and InceptionTime, have yet to be employed in the MTSC problems. In addition to this, Zhang et al. (2020) proposed a novel approach called TapNet—an attentional prototype network. This model attempted to combine the strengths of traditional and deep learning methods in solving MTSC tasks and successfully achieved high accuracy using the combination of the LSTM and random dimension permutation before convolution. Since TapNet can extract features from unlabelled (or shortage of labelled) data with little domain knowledge, we considered TapNet to be suitable for our dataset. This implementation gave us insight into adapting TapNet to an LSTM + FCN method. For the baseline LSTM, our implementation was inspired by the LSTM encoder-decoder model, stated by Buroni et al. (2021).

Buroni et al. (2021) implemented an LSTM encoder-decoder model to predict traffic flow in upcoming timeframes based on past observations. They utilised the same OBU data collected by Brussels Mobility from the 1st of January to the 28th of February 2019 (i.e., for Belgium, with 30 min temporal resolution). All of the data preprocessing and analysis was done using Pandas DataFrames. Time covariate variables were then generated from the date-time object to be used as features of the LSTM model. There were a total of four features: "Sine" and "Cosine" transformation of hours of the day, "day of the week" (which represents what day it is), and "working/weekend" days. These features were proven to be deterministic features that influence traffic. Therefore, the authors decided to use these features as the starting point and extended the idea as explained in Sect. 3.

Buroni et al. (2018) also offered a solution of Big Data platforms for better data collection, storing, and analysis activities. It was claimed to be especially effective for large streaming data. The solution involved a Hadoop Distributed File System (HDFS) to store data and a combination of Pandas DataFrame and PySpark for data reading, processing, and writing. Using Spark in their project was proven to improve the existing architecture—their solution was scalable and fast. Therefore, we decided to try out our new solution, using SparkSQL, instead of Pandas DataFrame, for all of our preprocessing and dataset operations.

3 Methodology

3.1 Data Preprocessing

Data preprocessing is a significant process in solving a big data problem because different actions taken in this process could affect the results of predictive models (Gonzalez Zelaya 2019). With appropriate preprocessing steps, raw data could be turned into high-quality data, and more meaningful information could be obtained. Data preprocessing included different steps such as data cleaning, integration, transformation, reduction, and discretisation (Cho and Choi 2017).

The data preprocessing step is the main contribution of this chapter, particularly for smart urban systems from the perspective of management and optimisation. We used the SparkSQL component available in Apache Spark, allowing us to do large-scale data processing using Spark DataFrames, instead of Pandas DataFrames used by Buroni et al. (2021). Spark DataFrames were created from the raw data of CSV format according to the defined schema (Table 1).

3.1.1 Data Cleaning and Aggregation

Due to a lack of data from 2020, the time series predictions for 2019 and 2021 were made separately. For the two periods of 2021 data, we appended the two datasets that belong to the same region, with the exact temporal resolution, into one dataset, without allowing duplication (using "union" and "distinct" operations). We noted that the first 2021 data ended on June 6th, while the second 2021 data started on the 5th, which means repetition on June 5th.

We then dropped all rows with missing values, grouped by the data based on the timestamp, and got the total number of trucks (from all street segments).

3.1.2 Feature Engineering

Feature engineering could be defined as a technique of creating new variables from the raw data in order to simplify data transformation and training while enhancing model accuracy (Patel 2023). Based on our data exploration (refer to next section), the "Datetime" column was converted into several columns that are more meaningful

Table 1 Spark DataFrame schema

Column name	Data type
Datetime	TimestampType
Street_ID	StringType
Traffic_Count	IntegerType
Traffic_Speed	FloatType

as features. This task was performed using functions available in PySpark SQL for Date and Timestamp data types. The features include:

- "DayOfWeek": day of the week (the corresponding day of that date, encoded as 1 = Sunday, 2 = Monday, 3 = Tuesday, 4 = Wednesday, 5 = Thursday, 6 = Friday, 7 = Saturday)
- "Date": date of the timestamp (from 1 to 31)
- "Month": month of the timestamp (from 1 to 12)
- "Year": year of the timestamp (2019 or 2021)
- "Hour": hour of the timestamp (from 0 to 23)
- "Minute": minute of the timestamp (from 0 to 59)
- "Second": second of the timestamp (0).

Next, we encoded some columns into categorical features using the PySpark SQL "when otherwise" functions:

- "Day_Or_Night": classifies if the timestamp is day or night. The "Day" category is from 6 a.m. to 6 p.m. and is encoded as 0, while the "Night" category is from 6 p.m. to 6 a.m. and is encoded as 1.
- "Weekdays_Or_Weekend": classifies if the timestamp is on a weekday (Monday–Friday), or at the weekend (Saturday–Sunday). The "weekday" category is encoded as 0, while the "weekend" category is encoded as 1.

With this method, the "Datetime" column was transformed into valuable features that could be used to predict the traffic count in future timestamps.

3.2 *Time Series Prediction Models*

3.2.1 Simple LSTM Model

Long Short Term Memory (LSTM) is one of the most popular deep learning models in recent years, which is both universal and effective in capturing long-term dependence (Greff et al. 2017). LSTM is widely used in time sequence prediction (Yadav et al. 2019). The OBU dataset has random noise, long-term trends, multi-features, and seasonal and cyclical fluctuations. Thus, we agreed that using the LSTM model for this multi-features time sequence dataset is very suitable.

The dataset was divided into 80% training data and 20% test data, which were used to fit the model and evaluate the accuracy of the model (Joseph and Vakayil 2021). A 5-fold cross-validation is also used to avoid over-fitting or under-fitting. After preprocessing, nine features were produced. We compared and evaluated the model performance, which used nine features against the model produced by Buroni et al. (2021), which used only four features. RobustScaler is used for data normalisation. The formula is as follows:

$$x' = \frac{x - x_{median}}{IQR} \tag{1}$$

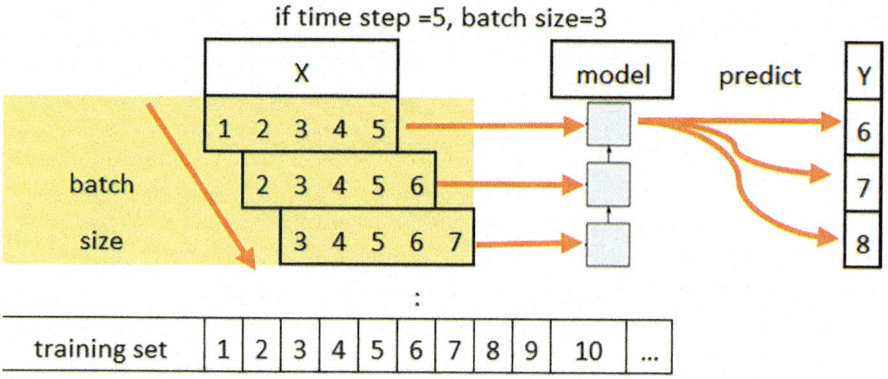

Fig. 1 An example of how multi-step LSTM model moved

Robust statistics are suitable for scaling data with abnormal values (departure). This scalar removes the median and scales the data according to the quantile range (the default is IQR or interquartile range). After normalisation, the training and test sets were reshaped so that the LSTM could accept them. The shape of the input dataset is (sample, time step, and features). LSTM always needs to predict the next value through the previous x (x = time step) data in multi-step time sequence models. This process is as shown in Fig. 1.

The time step was set to 24, and the datasets were divided into train X, train Y, test X, and test Y, which means that our model will predict the next value based on the first 24 data. The LSTM model was built using Keras based on TensorFlow 2.0. Since the size of the dataset was limited (as the smallest dataset size does not even exceed 5000 rows), we concluded that building a simpler LSTM model would have higher accuracy.

We first set the epoch to 200 times while adjusting the parameters and found that the model overfitted after the 120th time during the training process, meaning the best epoch should be around 120. At the same time, we tried to set the batch_size to 64, 32, and 16 and found that since the sample size is around 3000 and the epoch is around 120, setting the batch size to 32 can be more careful to train the model at a reasonable speed. The LSTM model includes:

- Bidirectional layer: packs LSTM into a bidirectional recurrent network to enhance feature extraction capabilities.
- Dropout layer: randomly disconnects input neurons at a rate = 0.3 every time the parameters are updated during the training process to prevent overfitting.
- Dense layer: set the output dimension to 1 (units = 1).

Figure 2 depicts the line chart of loss MAE (Mean Absolute Error) and accuracy MSE (Mean Squared Error) using the Belgium_30min_2019_4 features dataset.

After the prediction, the model is saved as an H5 file, and all prediction results are saved into PKL files.

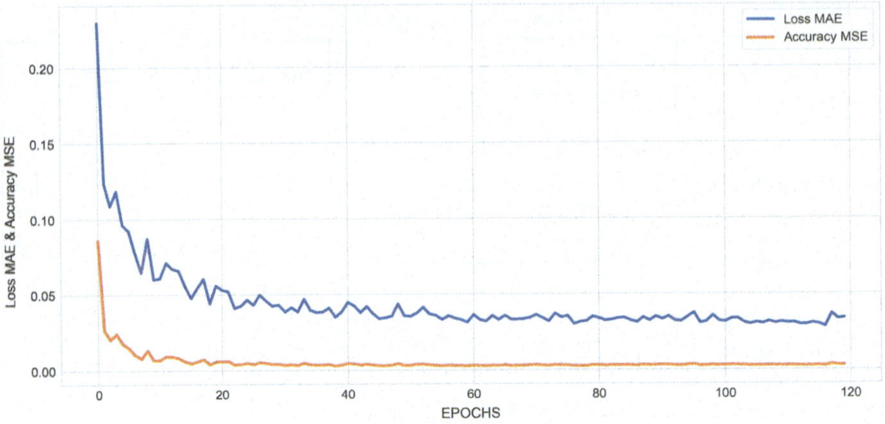

Fig. 2 Compare loss MAE and accuracy MSE value

3.2.2 LSTM + FCN Model

For comparison, LSTM + FCN models are also trained. We tried to extract more features at low-level features by FCN for comparison with the LSTM model. The data are processed by LSTM and FCN separately, and after being processed separately, the data are combined for the final output. This idea is inspired by TapNet (Zhang et al. 2020). The LSTM + FCN model is more suitable for our training data since the model is much smaller and easier to train using only thousands of training data.

In order to evaluate the performance of the model more accurately and objectively, three linear regression algorithm evaluation indicators are used. The formulas are as the following:

R^2 Score is used to measure the accuracy of the model (Eq. 2).

$$R^2 = 1 - \frac{\sum_i \left(\hat{y}^{(i)} - y^{(i)}\right)^2}{\sum_i \left(\bar{y}^{(i)} - y^{(i)}\right)^2} \tag{2}$$

Mean Absolute Error (MAE) (Eq. 3) and Root Mean Squared Error (RMSE) are used to calculate the errors (Eq. 4).

$$\frac{1}{m}\sum_{i=1}^{m} |y_{test}^{(i)} - \hat{y}_{test}^{(i)}| \tag{3}$$

$$\sqrt{\frac{1}{m}\sum_{i=1}^{m}\left(y_{test}^{(i)} - \hat{y}_{test}^{(i)}\right)^2} = \sqrt{MSE_{test}} \tag{4}$$

4 Results and Discussions

4.1 Data Exploration and Visualisation

Data exploration is the initial step used to understand patterns, trends, characteristics, and points of interest in big data problems before solving them (Bikakis et al. 2019). As the raw dataset is time-series data, we considered timestamp (in the "Datetime" column) as the only independent variable and "Traffic_Count" as the dependent variable we wanted to predict. However, the "Datetime" column is only a single column that does not give enough meaningful information.

Our exploration was based on several hypotheses listed as follows:

- The number of trucks varies daily; the difference will not only be during weekdays and weekends, but we are expecting a larger volume of trucks on a specific day.
- The number of trucks is higher during specific months, e.g., due to the holiday season.
- The number of trucks varies in different years, e.g., due to economic changes or global issues.
- Traffic is busier during the day, rather than at night.

With this motivation, figures that visualise the total number of trucks for each day of the week and hours for each day are generated using the Matplotlib library to understand the dataset better.

Figure 3 depicts the total number of trucks on each day of the week in the three different periods. The traffic is much higher on the weekdays, which agrees with the claim made by Buroni et al. (2021). In 2019, Thursday appeared to be the busiest day of the week, but in 2021, the weekday traffic was approximately the same. Both 2019 and 2021 data showed that Saturdays were busier than Sundays on weekends. The traffic was seen to be much higher in 2019, then dropped drastically in 2021, most probably due to the COVID-19 pandemic.

Figures 4, 5, and 6 focus on the average freight count on different hours daily. The number of trucks was high around 5–6 a.m.—possibly because trucks delivering goods early in the morning to avoid traffic jams—peaked around 11 p.m. This was possible because of working time in Belgium, i.e., from 6 a.m. to 8 p.m., and must not exceed 8 h a day (Belgium Federal Public Service 2023). So, if the driver starts around 5 a.m., he will finish working at 1 p.m. In all three figures, the freight traffic on weekends was low, as most businesses are closed during weekends. From all findings and analysis, we could imply that every feature we suggested affects freight traffic factors.

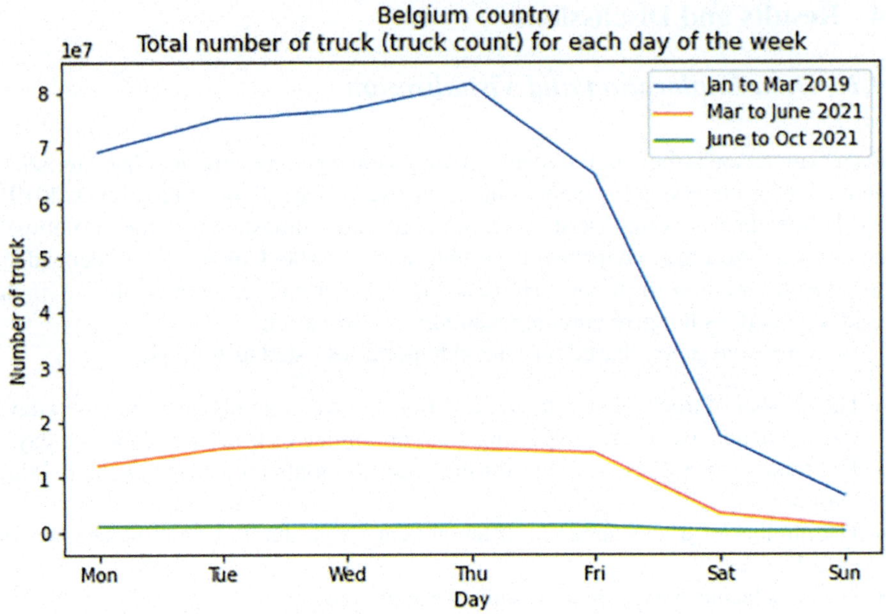

Fig. 3 Belgium total number of truck on each day of the week in 2019 and 2021

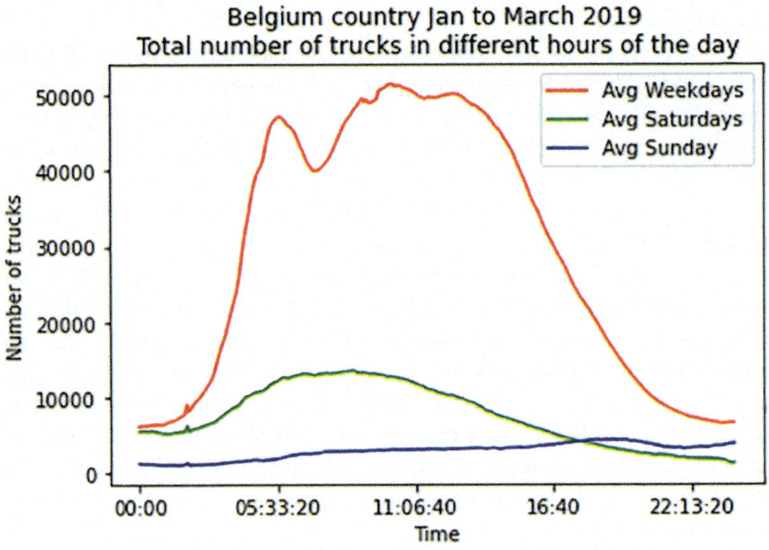

Fig. 4 Belgium total number of truck - day time (Jan to Mar 2019)

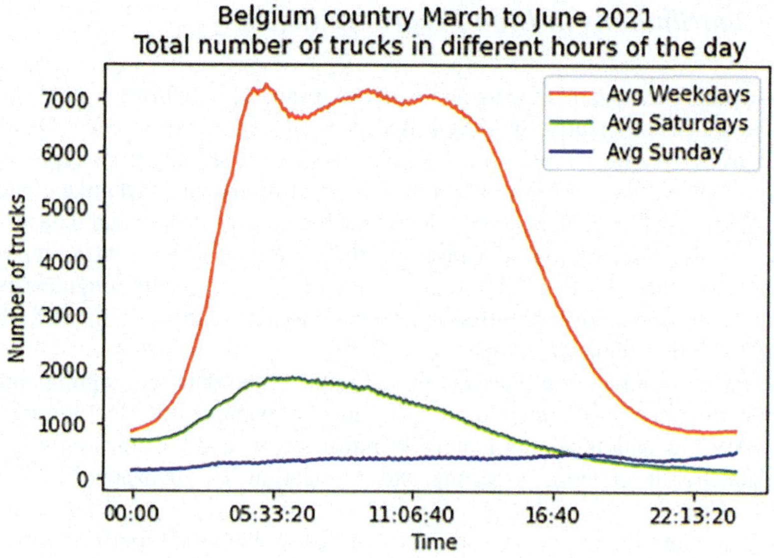

Fig. 5 Belgium total number of truck - day time (Mar to Jun 2021)

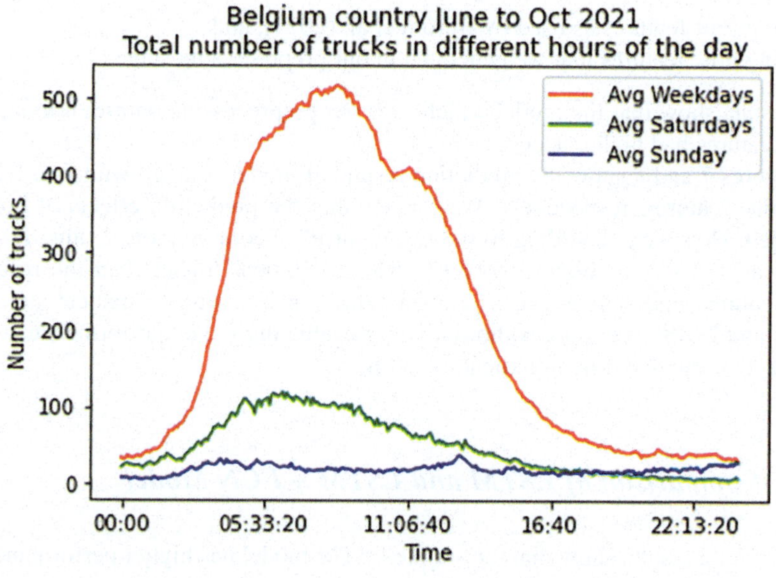

Fig. 6 Belgium total number of truck - day time (Jun to Oct 2021)

4.2 Contributions in Data Preprocessing

The data preprocessing method using Spark DataFrames proved very suitable for handling our dataset efficiently. SparkSQL interface allowed users to create a DataFrame directly from CSV files. It provided information such as data structure, types, dimensions, and the number of rows to optimise the operations with SQL-like commands better. With SparkSQL, users could define each column's designated data structure and types using data schema. For instance, the 'Datetime' data, a String type in the CSV file, was turned into a 'Timestamp' type, making the feature engineering step simpler, faster, and more straightforward because we could utilise PySpark SQL Date and built-in Timestamp functions.

The group-by operation that we applied reduced the data size significantly and made the time prediction much more efficient. For example, the Bel_30min_0101_0103_2019 raw data is approximately 11 million rows and becomes only less than 3000 data after the cleaning, grouping, and aggregation. This helps our LSTM model to train faster.

We compared the impact of different features on the model performance, using the same dataset and the same model parameters on the baseline LSTM. We trained two models respectively:

- Using four features as used by Buroni et al. (2021), and
- Using nine features that we generated in our preprocessing stage.

Results show that the model which uses our preprocessing dataset training with nine features had higher accuracy.

Figures 7 and 8 show the prediction results of models trained with four features and nine features, respectively. We can see that the prediction effects of both are satisfied. However, according to the results of R2_Score in Table 2, nine features' result accuracy is as high as 99.89%, which is 0.06% higher than the results of four features, and there are also significant differences in other statistical indicators. MAE and RMSE are more authoritative for evaluating model performance, and we represent these two critical indicators in Fig. 9.

4.3 Comparison of LSTM and LSTM + FCN Model

The training results show that the simple LSTM model has higher performance and lower MAE and RMSE values.

Figure 10 shows the performance of the LSTM + FCN and LSTM models for four features and nine features, respectively. The figure shows that the LSTM + FCN performance is not as good as the LSTM model. In addition, as before, we can see that the results for the case with four features as training input are lower than the results of the nine features we generated.

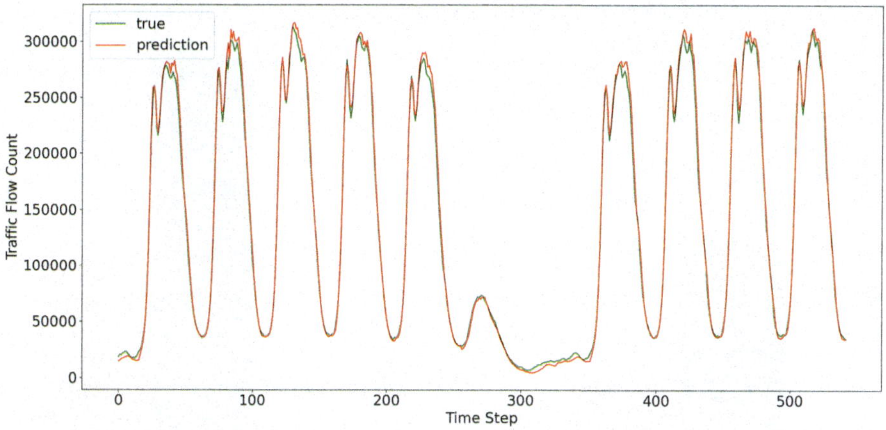

Fig. 7 The prediction results use a model trained by the dataset Belgium 30 min from 2019/01/01 to 2019/03/01 with four features by Buroni et al. (2021)

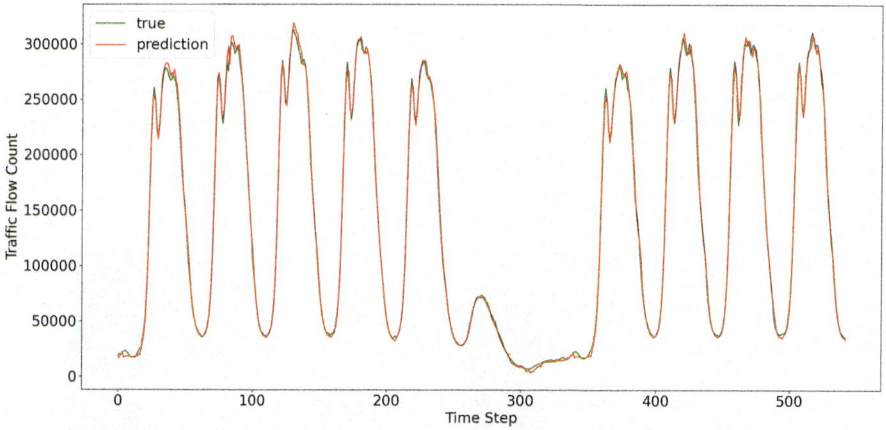

Fig. 8 The prediction results use a model trained by the dataset Belgium 30 min from 2019/01/01 to 2019/03/01 with nine features

Table 2 Compare the model trained based on four and nine-feature datasets

Name	R2_Score	MAE	MSE	RMSE
4 features	0.99836	3055.68	18877004.34	4344.76
9 features	0.99890	2590.46	12666090.95	3558.94

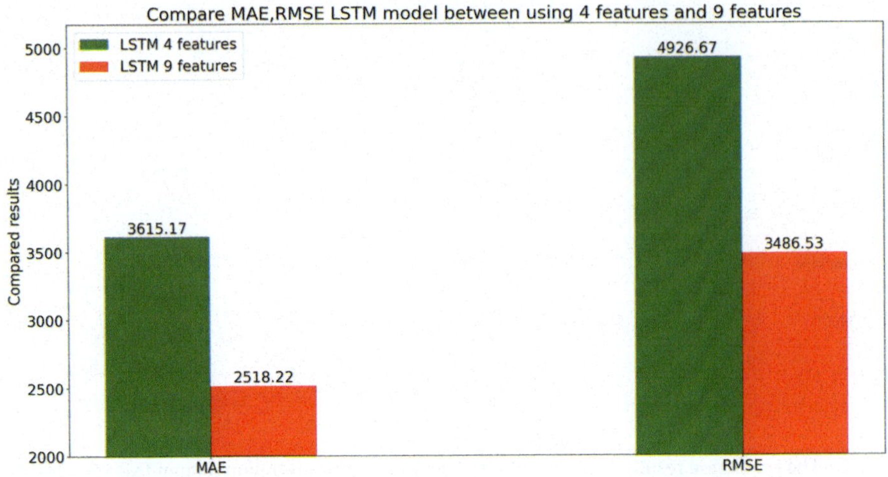

Fig. 9 Comparison of LSTM model using four and nine features

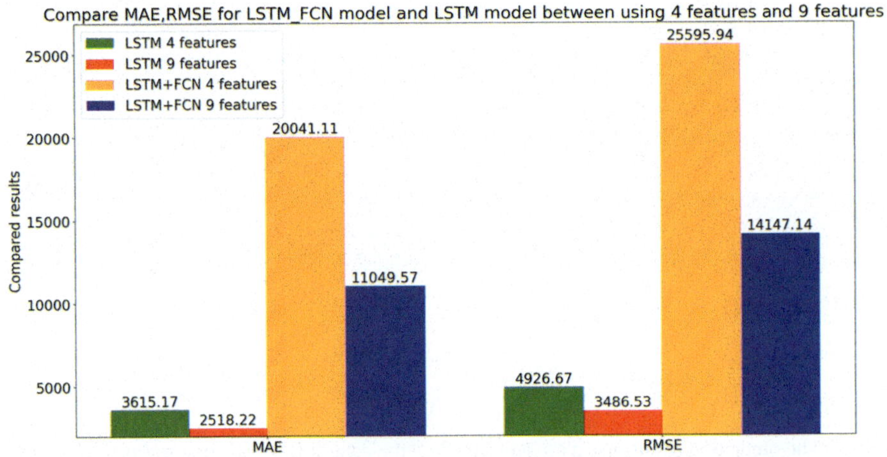

Fig. 10 Comparison of LSTM and LSTM + FCN model using four and nine features

This result is because the LSTM model is sufficient to capture the sequential features, but the LSTM + FCN model is too complex compared to the LSTM. The LSTM + FCN model introduces several additional convolutional layers, which leads to the model not being able to learn enough low-latitude features when the amount of data is low, as shown in Fig. 11. In contrast, we can see in Fig. 8 that the LSTM model learns these features sufficiently at a time of about 300.

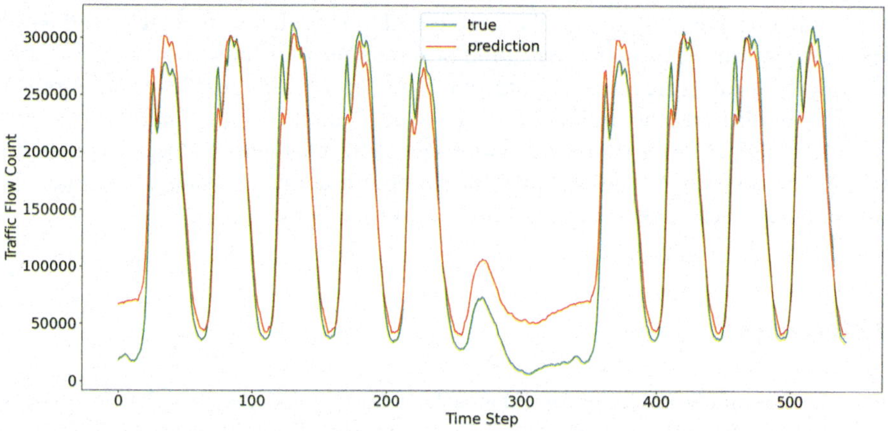

Fig. 11 The prediction results use an LSTM + FCN model trained by the dataset Belgium 30 min from 2019/01/01 to 2019/0301 with nine features

5 Conclusion

This study is based on analysing time-series data obtained from On-Board Units installed on Belgium freight vehicles and making predictions in future timestamps based on past data and patterns. Big data analysis is used to evaluate a particular urban system in a particular context. Backgrounds and motivations behind analysing freight transport data are presented, indicating the relevance and details of available data used in this study. The predictive models proposed in this chapter could forecast and understand freight traffic at a particular time—i.e., giving valuable insights to minimise adverse effects of freight transport and apply better traffic policies in Belgium. In this chapter, we reviewed existing studies on this topic and implemented the LSTM and LSTM + FCN models trained with the proposed extracted features. We contributed to solving big data problems related to the quantity of data and speed of operations. With our novel solution, preprocessing the raw data using Spark DataFrames, data cleaning, aggregation, and feature engineering was more straightforward and fast. We performed data exploration and generated nine features from the datetime variable, which improved the existing model accuracy further.

Our second contribution was to the LSTM deep learning model for prediction. We first built a simple LSTM model with three layers based on the size of the preprocessed dataset and adjusted the most suitable hyper-parameters to get the most accurate results. Our simple model has higher prediction accuracy than the LSTM + FCN model for all Belgium 30-min dataset, reaching an accuracy of 99.89%. This shows that our proposed approach is well-fitted for the dataset and highly optimised to give accurate predictions.

With these results, we have highlighted the importance of data exploration and preprocessing stages and how a simple model could be well-suited for a specific multi-variate time series classification task. Although the final LSTM + FCN model we worked on is not as good as the simple LSTM model, this attempt is well worth mentioning. This implementation could be a reference for future research and other scholars whose data are more suitable for this model. Thus, there is scope for improvement by minimising the limitations of this study's model. In the future, the preprocessing and LSTM could be further optimised to lower the running speed and memory usage. In this way, we could consider further optimisation of urban systems and their operations, such as larger transportation network systems.

References

Bagnall, Anthony, Michael Flynn, James Large, Jason Lines, and Matthew Middlehurst. 2020. On the usage and performance of the hierarchical vote collective of transformation-based ensembles version 1.0 (hive-cote v1. 0). In *International Workshop on Advanced Analytics and Learning on Temporal Data*, 3–18. Springer.

Ballantyne, Erica E.F., Maria Lindholm, and Anthony Whiteing. 2013. A comparative study of urban freight transport planning: Addressing stakeholder needs. *Journal of Transport Geography* 32: 93–101. https://doi.org/10.1016/j.jtrangeo.2013.08.013.

Beeram, Satyanarayana, and Swarna Kuchibhotla. 2020. *Time series analysis on univariate and multivariate variables: A comprehensive survey*, 119–126. ISBN: 978-981-15-5396-7. https://doi.org/10.1007/978-981-15-5397-4_13.

Belgium Federal Public Service. 2023. Working time and rest periods. https://employment.belgium.be/en/themes/international/posting/working-conditions-be-respected-case-posting-belgium/working-time-and. Retrieved February 2023.

Bikakis, Nikos, George Papastefanatos, and Olga Papaemmanouil. 2019. Big data exploration, visualization and analytics. *Big Data Research* 18 (1).

Buroni, Giovanni, Gianluca Bontempi, and Karl Determe. 2021. A tutorial on network-wide multi-horizon traffic forecasting with deep learning. In *EDBT/ICDT Workshops*.

Buroni, Giovanni, Yann-Ael Le Borgne, Gianluca Bontempi, and Karl Determe. 2018. On-board-unit data: A big data platform for scalable storage and processing. In *2018 4th International Conference on Cloud Computing Technologies and Applications (Cloudtech)*, 1–5. IEEE.

Cho, Wonhee, and Eunmi Choi. 2017. Big data pre-processing methods with vehicle driving data using MapReduce techniques. *The Journal of Supercomputing* 73 (7): 3179–3195.

Dempster, Angus, François Petitjean, and Geoffrey I. Webb. 2020. Rocket: Exceptionally fast and accurate time series classification using random convolutional kernels. *Data Mining and Knowledge Discovery* 34 (5): 1454–1495.

Deng, Houtao, George Runger, Eugene Tuv, and Martyanov Vladimir. 2013. A time series forest for classification and feature extraction. *Information Sciences* 239: 142–153.

Gonzalez Zelaya, Carlos Vladimiro. 2019. Towards explaining the effects of data preprocessing on machine learning. In *2019 IEEE 35th International Conference on Data Engineering (ICDE)*, 2086–2090. https://doi.org/10.1109/ICDE.2019.00245.

Greff, Klaus, Rupesh K. Srivastava, Jan Koutník, Bas R. Steunebrink, and Jürgen Schmidhuber. 2017. LSTM: A search space odyssey. *IEEE Transactions on Neural Networks and Learning Systems* 28 (10): 2222–2232. https://doi.org/10.1109/TNNLS.2016.2582924.

He, Kaiming, Xiangyu Zhang, Shaoqing Ren, and Jian Sun. 2016. Deep residual learning for image recognition. In *Proceedings of the IEEE Conference on Computer Vision and Pattern Recognition*, 770–778.

Ismail Fawaz, Hassan, Benjamin Lucas, Germain Forestier, Charlotte Pelletier, Daniel F. Schmidt, Jonathan Weber, Geoffrey I. Webb, Lhassane Idoumghar, Pierre-Alain Muller, and François Petitjean. 2020. InceptionTime: Finding AlexNet for time series classification. *Data Mining and Knowledge Discovery* 34 (6): 1936–1962.

Ismail Fawaz, Hassan, Germain Forestier, Jonathan Weber, Lhassane Idoumghar, and Pierre-Alain Muller. 2019. Deep learning for time series classification: A review. *Data Mining and Knowledge Discovery* 33 (4): 917–963.

Joseph, V. Roshan, and Akhil Vakayil. 2021. Split: An optimal method for data splitting. *Technometrics* 1–11.

Karim, Fazle, Somshubra Majumdar, Houshang Darabi, and Samuel Harford. 2019. Multivariate LSTM-FCNS for time series classification. *Neural Networks* 116: 237–245. https://doi.org/10.1016/j.neunet.2019.04.014.

Karlsson, Isak, Panagiotis Papapetrou, and Henrik Boström. 2016. Generalized random shapelet forests. *Data Mining and Knowledge Discovery* 30 (5): 1053–1085.

Le Nguyen, Thach, Severin Gsponer, and Georgiana Ifrim. 2017. Time series classification by sequence learning in all-subsequence space. In *2017 IEEE 33rd International Conference on Data Engineering (ICDE)*, 947–958. IEEE.

Le Nguyen, Thach, Severin Gsponer, Iulia Ilie, Martin O'Reilly, and Georgiana Ifrim. 2019. Interpretable time series classification using linear models and multi-resolution multi-domain symbolic representations. *Data Mining and Knowledge Discovery* 33 (4): 1183–1222.

Li, Lincan, Chiew Foong Kwong, Qianyu Liu, Pushpendu Kar, and Saeid Pourroostaei Ardakani. 2021. A novel cooperative cache policy for wireless networks. *Wireless Communications and Mobile Computing* 2021: 1–18. https://doi.org/10.1155/2021/5568935.

Lubba, Carl H., Sarab S. Sethi, Philip Knaute, Simon R. Schultz, Ben D. Fulcher, and Nick S. Jones. 2019. catch22: Canonical time-series characteristics. *Data Mining and Knowledge Discovery* 33 (6): 1821–1852.

Middlehurst, Matthew, James Large, and Anthony Bagnall. 2020. The canonical interval forest (CIF) classifier for time series classification. In *2020 IEEE International Conference on Big Data (Big Data)*, 188–195. IEEE.

Patel, Harshil. 2023. What is feature engineering-importance, tools and techniques for machine learning. https://towardsdatascience.com/what-is-feature-engineering-importance-tools-and-techniques-for-machine-learning-2080b0269f10. Retrieved March 2023.

Sadeghi-Niaraki, Abolghasem, Masood Varshosaz, Kyehyun Kim, and Jason J. Jung. 2011. Real world representation of a road network for route planning in GIS. *Expert Systems with Applications* 38 (10): 11999–12008. https://doi.org/10.1016/j.eswa.2010.12.123.

Schäfer, Patrick, and Ulf Leser. 2017. Multivariate time series classification with WEASEL+MUSE. arXiv preprint arXiv:1711.11343.

Shokoohi-Yekta, Mohammad, Hu Bing, Hongxia Jin, Jun Wang, and Eamonn Keogh. 2017. Generalizing DTW to the multi-dimensional case requires an adaptive approach. *Data Mining and Knowledge Discovery* 31 (1): 1–31.

Szegedy, Christian, Sergey Ioffe, Vincent Vanhoucke, and Alex Alemi. 2016. Inception-v4, Inception-ResNet and the impact of residual connections on learning.

ULB. 2023. Brussels mobi-aid: Brussels mobility advanced indicators dashboard. https://mlg.ulb.ac.be/wordpress/portfoliopage/brussels-mobi-aid-brussels-mobility-advanced-indicators-dashboard/. Retrieved March 2023.

Wang, Zhiguang, Weizhong Yan, and Tim Oates. 2017. Time series classification from scratch with deep neural networks: A strong baseline. In *2017 International Joint Conference on Neural Networks (IJCNN)*, 1578–1585. IEEE.

Yadav, Anita, C.K. Jha, and Aditi Sharan. 2019. Optimizing LSTM for time series prediction in Indian stock market. In *International Conference on Computational Intelligence and Data Science*, India.

Zhang, Xuchao, Yifeng Gao, Jessica Lin, and Chang-Tien Lu. 2020. TapNet: Multivariate time series classification with attentional prototypical network. In *Proceedings of the AAAI Conference on Artificial Intelligence*, vol. 34, 6845–6852.

Chapter 4
Data-Driven Multi-target Prediction Analysis for Driving Pattern Recognition: A Machine Learning Approach to Enhance Prediction Accuracy

Abstract The driving pattern is critical in life quality enhancement, road traffic minimisation, and transportation risk reduction. It includes several parameters, such as the car's speed, traffic, weather, and road status. This article investigates the correlations between driving attributes and proposes a multi-target prediction model to recognise driving patterns. For this, a pre-processing data approach, including Pearson's Correlation Coefficient, Quantile Discretisation, and Hybrid Data Resampling, is used to reduce data feature dimensions and remove meaningless variables. The Random Forest technique predicts four targets, including vehicle speed, rain intensity, driver's well-being, and driver's rush. At the same time, the K-Nearest Neighbours algorithm (KNN) groups the prediction results and forms the driving patterns. The performance of the proposed multi-target prediction model is compared with four classifiers, including Multilayer Perceptron, Decision Tree, Multinomial Logistics Regression, and LR one versus rest. According to the results, the Random Forest model outperforms the benchmarks regarding prediction accuracy. The findings of this chapter help optimise prediction accuracy that could then be used for urban transportation system optimisation.

Keywords Driving pattern recognition · Multi-target prediction · Road traffic · KNN · Pearson correlation

1 Introduction

Driving patterns are usually used to model driver's behaviour, resource consumption (e.g., oil), traffic condition, and road safety. They are highly correlated with the vehicle's speed, weather conditions, journey time, and traversed distance (Yu 2012). For this, driving pattern recognition has been extensively investigated in several studies, especially in the field of transportation systems since the 1970s (Watson 1978). Figure 1 summarises the key factors which are usually used to address driving patterns (Ericsson 2000):

The research work in this chapter was supported by our research team members: Youtian Peng, Dongchi Li, Wenyue Zheng, and Ke Xu.

S. Pourroostaei Ardakani and A. Cheshmehzangi, *Big Data Analytics for Smart Transport and Healthcare Systems*, Urban Sustainability,
https://doi.org/10.1007/978-981-99-6620-2_4

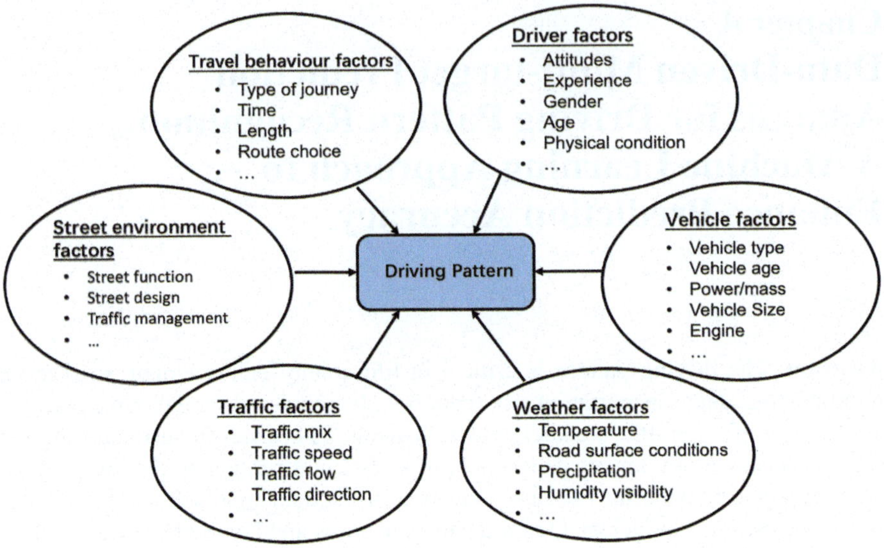

Fig. 1 Driving pattern's factors (redrawn by the authors, and adapted from Ericsson (2000))

1. Street environment factors such as speed limit, street design, and street function.
2. Traffic factors—mainly traffic flow condition, and vehicle number.
3. Weather factors include rain/snow intensity, humidity, and visibility.
4. Driver factors—mainly driver's skill, mood, gender, and age.
5. Vehicle factors such as vehicle type, engine RPM, and car's load.
6. Travel behaviour factors including journey type and route choice.

Driving pattern recognition has the capacity to offer several benefits, such as transportation planning, safety, economy, resource conservation, and/or environment engineering (Ericsson 2001). There are multiple benefits if driving patterns are optimised. More importantly, it can be used to study and predict driver behaviours (e.g., well-being), transportation factors (e.g., traffic), and environmental parameters (e.g., air quality) (Elgowainy et al. 2010; Neubauer et al. 2013). All these parameters are vital to create healthy and sustainable cities and communities. For example, Lin et al. (2004) used the results of dividing pattern analysis to predict vehicle emission, whereas Ardakani et al. (2022) studied the impact of road driving patterns and regional transportation lock down on COVID-19 spread in Ningbo, China. In conducting such studies, researchers and scientists have become increasingly interested in transportation data analysis (i.e., road, air, sea, and rail) to recognise driving patterns and optimise transportation plans. They utilise machine learning and Big data analysis techniques to analyse large datasets, extract meaningful data features, and train high-quality predictive models. Such approaches indicate high-level optimisation opportunities at larger scales of cities, regions, between regions, as well as inter-city and intra-city transportation networks.

Machine learning (ML) prediction models may need to take the benefits of Big data analysis approaches if the datasets are huge and change quickly. ML classification techniques can be categorised into two classes in terms of target variety: single and multiple targets. Single-target prediction aims to predict only one variable according to the predictor features (e.g., driving, vehicle, weather, and travel conditions), whereas multi-target prediction proposes a model to predict several targets based on the given dataset. There are two methods to build multi-target prediction models: local and global (Zhou et al. 2004; Kocev et al. 2013). Local multi-target prediction is usually used to address prediction applications with a limited number of targets. It selects the most relevant features from the original feature space and trains a predictive model for each target. Local multi-target prediction avoids taking all the features into account at once, but it finds the best-fitted model for each target. On the other hand, global multi-target prediction uses only one model to predict multiple targets at once. It has the capacity to reduce model training delay and enhance the model generalisation as compared to the local multi-target prediction (Last 2004).

This research study uses a large public dataset (i.e., Car-trip; Vitor 2023) to propose a global multi-target classification model for driving pattern prediction. As the dataset is huge, Apache Spark data analytic engine and MLlib libraries (Apache Spark 2023a) are used to implement data processing and machine learning approaches. A data pre-processing approach is used to extract data features, measure correlations, remove meaningless variables, discretise data samples, and balance the dataset. The Random Forest technique (Cutler et al. 2012) builds a multi-target prediction model, and the K-Nearest Neighbours algorithm (KNN) (Wagstaff et al. 2001) groups the prediction results to recognise the driving patterns. The performance of the Random Forest (RF) model is measured and compared with four classification techniques, including Multilayer Perceptron (Ramchoun et al. 2016), Decision Tree (Song and Ying 2015), Multinomial Logistics Regression (Starkweather and Moske 2011), and Logistics Regression-based one versus rest classifiers (Tax and Duin 2002). The key contributions of this research are outlined below:

- Use a Big data pre-processing approach to prepare a massive public transportation dataset for machine learning model training.
- To propose a global multi-target prediction model with minimised training delay.
- To group the prediction results and form the driving patterns.
- To evaluate and compare the performance of the proposed predictive ML model with four benchmarks using an extensive data-driven approach.

The rest of this chapter is organised as follows. Section 2 introduces driving patterns and reports a review of machine learning techniques for driving pattern recognition and analysis. Section 3 describes the research methodology, including data pre-processing, multi-target classification model, and driving pattern recognition. Section 4 outlines the experimental plan and discusses the results. Section 5 concludes the critical points of this research and highlights the issues that should be addressed as further work.

2 Literature Review

This section introduces driving patterns, reviews state-of-the-art literature in this field of research, and outlines machine learning techniques for multi-target prediction and driving pattern recognition.

A driving pattern is usually defined as the vehicle's speed and/or driving behaviour (i.e., shifting and braking). Ericsson (2000) outlines driving speed as one of the key metrics in driving pattern recognition, while Bella (2014) uses the CRISS driving simulator to analyse the impact of speed on the driving pattern. As they report, driving speed is the key factor that is influenced by driving behaviours, environmental conditions, and driver feelings during transportation. In addition, driving patterns can be influenced by other factors such as environment, traffic, and the driver. According to Kuhler and Karstens (1978), speed, average acceleration, average driving cycle length, and vehicle stationary time ratio are the most common driving pattern factors. Moreover, Andre (1996) includes a combination of vehicle stops per kilometre, average speed, positive kinetic energy, parking period, acceleration, and deceleration as the driving pattern. Wang et al. (2018) use driving behaviours metrics (i.e., decision-making) to define the driving patterns, whereas Higgs and Abbas (2014) analyse driving behaviours and environment conditions (i.e., traffic, obstacle status, road width) to recognise driving patterns.

Machine learning techniques have the capacity to be used for driving pattern recognition and classification. They have the capacity to explore data correlations and predict/classify driving patterns. ML classification/prediction techniques are formed into four categories as below:

1. Binary classification: It aims to classify only one label, which contains binary values.
2. Multi-class classification: It classifies one label that may have more than two values.
3. Multi-label classification: The classifier addresses more than one label, but the values are binary.
4. Multi-target classification: This proposes a classification model with more than one label, each of which has more than two values. According to Kocev et al. (2013), a multi-target prediction is defined as follows:

   - A feature space F that may consist discrete or continuous features, i.e., $\forall F_i \in F, F_i = (f_{i_1}, f_{i_2}, \ldots, f_{i_S})$, where S is the size of feature vector.
   - A target space T that includes a vector of target variables, i.e., $\forall T_i \in T$, $T_i = (t_{i_1}, t_{i_2}, \ldots, t_{i_D})$, where D is the size of target vector.
   - A sample set X, each sample is a tuple of the feature and target space, i.e., $\forall X = (F_i, T_i | F_i) \in F, T_i \in T, 1 \leq i \geq K$, K is the number of examples in $X(K = |X|)$.
   - A evaluation metric e which measures the performance of the model.

- A function $g : F \rightarrow T$ in which g should maximize e. g represents a machine learning algorithm (e.g., Decision Tree) to predict targets T from features F. It is a multi-target regression if t is a continuous variable, whereas it addresses a multi-target classification if t is a discrete variable.

Multi-target classification techniques are used to analyse data features and predict multiple targets. Kocev et al. (2007) combine the predictions of multiple classifiers to address a multi-target prediction. For this, two ensemble learning techniques including Bagging and the Random Forest are utilised to form an ensemble Decision Tree for multi-target classification. According to the results, the proposed multi-target Decision Tree outperforms the standard multi-target Decision Tree in terms of model training delay. Last (2004) introduces a commonly-used multi-objective info-fuzzy network (M-IFN) to build a multi-target classification model. It forms a graph and annotates the nodes with the same feature. M-IFN is formed by one root and a number of leaf nodes, each of which is connected to at least one target. Last et al. (2008) use M-IFN to build a multi-target model and predict 27 features, whereas Last et al. (2016) utilise M-IFN to predict six targets. According to the results, higher accuracy is achieved in multi-target prediction as compared with single-target prediction. Spanos and Angelis (2018) build three classification models, including Boosting, Random Forest, and Decision Tree. As it reports, the Random Forest outperforms Boosting and Decision Trees in terms of model accuracy. Schneck et al. (2017) take two Random Forest algorithms for multi-target classification. It proposes a solution to combine classes with skewed distribution to address multi-target classification. Levatić et al. (2018) build predictive clustering trees for multi-target prediction, that ensembles semi-supervised learning. It claims that the proposed method is interpretable, fast to learn, easy to understand, and able to deal with different feature types. Al Muktadir et al. (2019) propose a Majority-Voting case Ensemble algorithm (MVEN) to support multi-target prediction. It explains that MVEN addresses a better quality of service and reduces model training delays.

Several pieces of literature focus on machine learning model training in driving pattern analysis applications. Wang and Xi (2016) propose a K-mean clustering based on the Support Vector Machine technique (k-SVM) to classify driving behaviours and group them into two clusters, including aggressive or gentle. It reduces model training delay and improves the classification accuracy (i.e., 80%) as compared to SVM. However, it still lacks a multi-target prediction model to classify the driving patterns. To resolve it, Zhang et al. (2019) proposes a deep learning framework to build a multi-target prediction model. It predicts the behaviour of network-BUS (CAN-BUS) by analysing the onboard control data samples. Bernardi et al. (2018) uses a time series classification method to analyse driving behavioural parameters captured by the Automobile Monitoring System (AMS) and builds a model to identify the driving patterns. Choi et al. (2007) utilise a Hidden Markov Model to study the driver behaviour according to several parameters such as braking, speed, and acceleration. The objective is to classify vehicle mobility and analyse driver behaviours. However, it only focuses on driver behaviours and fails to include other factors such as weather, road status, and vehicle condition. Malik et al. (2021) proposes an ANN deep learning model to study and predict driving patterns. It processes and analyses a real-time dataset that is collected from the vehicle's Onboard Diagnostics (OBD) port for driving pattern classification. As this research reports, the ANN model is able to classify the diving behaviours with 99% accuracy. However, it avoids using environmental and driver data features and only utilises OBD data to predict and classify driving patterns. Ardakani et al. (2023) uses four machine learning techniques including decision trees, random forest, multinomial logistic regression, and naïve Bayes to explore road car accident data patterns. It proposes a predictive model by investigating meaningful data features, such as accident severity, the number of casualties, and the number of vehicles.

According to the literature review, a number of machine learning techniques (e.g., Random Forest, Logistic Regression, and Multilayer Reception) have been used to build predictive models in transportation applications and/or recognise the driving patterns (e.g., K-means and SVM). However, there still is a lack of extensive data-driven studies to analyse big transportation datasets and extract a wide range of transportation data features (e.g., environmental conditions, driver behaviours, and vehicle factors) to build a high-quality global multi-target prediction model for driving pattern recognition. The performance of the proposed model should be evaluated and compared with well-known classifiers to highlight the key findings and superiorities of this research (Fig. 2).

3 Methodology

This section describes the research methodology to build a global multi-target model for driving pattern prediction. It includes data pre-processing, multi-target prediction model training, and driving pattern recognition. As Fig. 3 shows, the data pre-processing approach includes attribute selection, discretisation, and data resampling techniques. The Random Forest technique is used to build the multi-target prediction, whereas the KNN clustering collects the prediction results and groups them as driving patterns.

3.1 Data Pre-processing

Raw datasets may include noise, meaningless, null, and/or missing values that usually result in reduced prediction accuracy and model training performance. For this, a data pre-processing approach is usually used to prepare the dataset for model training. However, traditional pre-processing approaches fail, and/or the data processing delay sharply increases if the dataset (i.e., Car-trip) is massive. Apache Spark (2023a) provides a data processing engine for large-scale datasets. It is used in this research study to support data pre-processing tasks (i.e., attribute selection, discretisation, and data resampling).

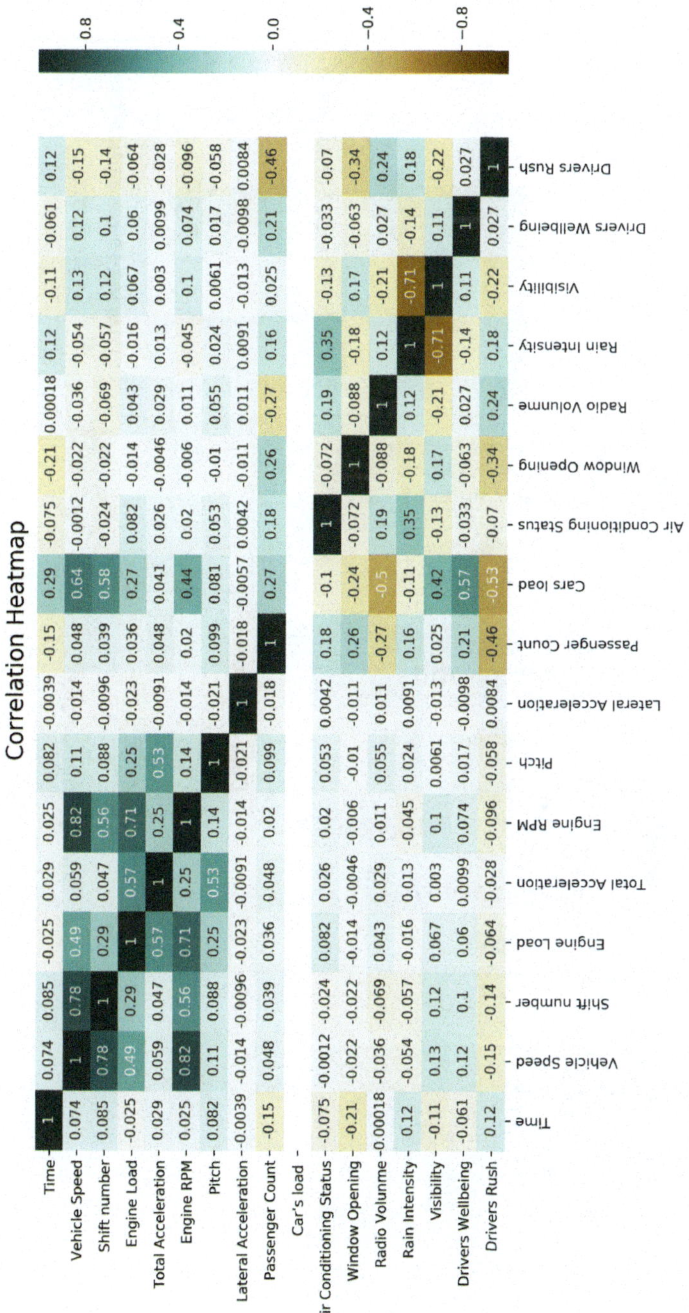

Fig. 2 The Pearson's correlation heat map

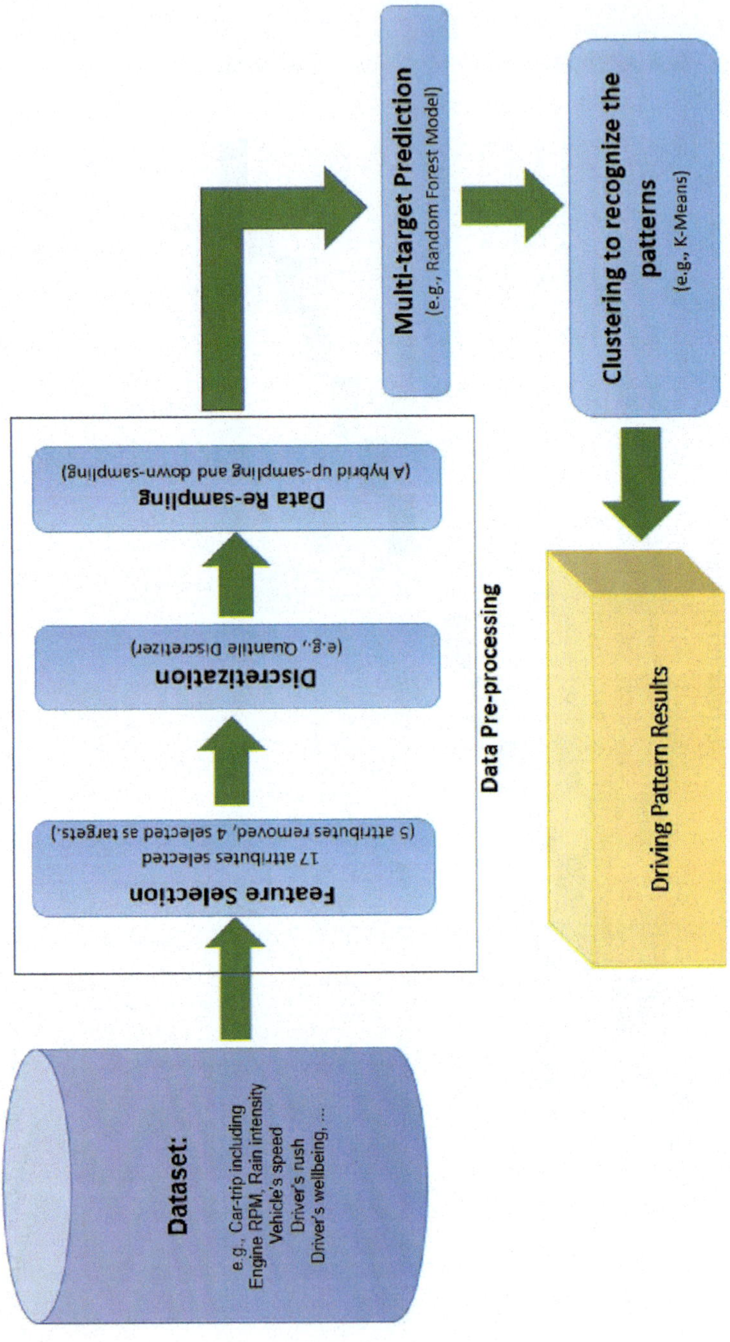

Fig. 3 The research's methodology diagram

Table 1 The data features

Feature	Description
Time	Time vector in one-hundredth seconds
Vehicle's speed	Car's speed in m/s
Shift number	A shift value reported by sensors (0 = intermediate position)
Engine load	Collected from engine load (in % of max power)
Total acceleration	Driving acceleration, collected from sensors (m/s^2)
Engine RPM	Reported by OBD sensor
Pitch	Vehicle's weight shift calculated by accelerometer data (XYZ)
Lateral acceleration	Driving acceleration around a curve, captured by sensors (m/s^2)
Passenger count	Reported by the driver (0–5) to show the number of passengers
Car's load	Vehicle's weight captured by a sensor
Air conditioning status	Collected from the vehicle's sensors
Window opening	Captured by a window sensor to indicate the windows' status
Radio volume	Collected from the sound sensors
Rain intensity	Collected from the vehicle's rain sensor
Visibility	Reported by the driver to indicate the road visibility (0–10)
Driver's wellbeing	Reported by the driver to indicate feeling (0–10)
Driver's rush	Reported by the driver to show rush driving (0–10)

3.1.1 Dataset

This study uses an original public dataset (Car-trip; Vitor 2023) that contains 7,748,753 records with 17 features. The data samples have been collected from 38 drivers with an average of 30–60 min of driving. The dataset is randomly divided into two parts, including 76% (29 drivers) to train and 24% (9 drivers) to test the machine learning model. The data features are reported by using vehicle sensory devices (e.g., rain intensity, visibility, engine RPM and windows opening) or the drivers (e.g., driver's well-being and rush) during transportation. Table 1 briefly introduces the dataset's features.

3.1.2 Feature Selection

The Pearson technique is used to measure and analyse the correlation coefficients between the data features. According to Benesty et al. (2009), two features have a minimum dependency if their correlation coefficient is above 0.3. As Fig. 2 shows, there are four attributes, including vehicle's speed, diver well-being, driver rush, and rain intensity that are highly correlated with others. They are selected as the targets. Moreover, there are eight features including road visibility, radio volume, window opening, air conditioning status, passenger count, shift, engine RPM, and engine load, are selected as predictors/features because they have the minimum correlation with the targets. However, the results indicate that five features, including pitch, time, car's load, lateral acceleration, and total acceleration have low correlations. They have no impact on prediction and should be removed.

3.1.3 Discretisation

Multi-target classification needs discretisation because it is unable to train discrete and continuous data at the same time (Al-Ibrahim 2011). This means that continuous variables should be discretised and form a discrete vector for model training. For this, *Quantile Discretiser* library under *pyspark.ml.feature* (Apache Spark 2023b) is used to support discretisation.

3.1.4 Data Resampling

Resampling techniques are usually used to adjust the size of data samples to avoid imbalanced data analysis, especially if the dataset is large. Resampling techniques are classified into two categories: up-sampling and down-sampling. The former (i.e., random copying) is often used to increase the size of minority classes when the number of data samples in each class is different. In contrast, the latter decreases the size of the majority classes. Up-sampling addresses overfitting if the number of data samples varies, and down-sampling results in information lost in big datasets.

This chapter utilises a hybrid resampling approach as a combination of up and down sampling. For this, Synthetic Minority Oversampling Technique (SMOTE) (Fernández et al. 2018) method is used to support oversampling. It avoids simple replication of minority class instances but updates the dataset by SMOTE synthetic samples within a defined neighbourhood. In addition, a down-sampling is used through random sample selection to reduce the size of majority classes.

3.2 Multi-target Prediction Model

The Random Forest technique is used in this article to build a multi-target classification model. It is a commonly used ensemble learning approach that has the capacity to support global multi-target prediction. Global multi-target prediction models have the capacity to increase model robustness and code maintainability as compared to local ones.

Algorithm 1 predicts a vector of labels (i.e., O) according to four target variables, including vehicle speed, rain intensity, driver's well-being, and driver's rush. It works based on two basic algorithms in the field of ensemble learning, including Bootstrap and Bagging. The former is a sampling method that refers to random sampling with put-back where each sample has the same probability of being drawn (Horowitz 2001). The samples are drawn and put back one by one until all the required samples are selected. The latter draws the new training samples from the original dataset to build a model for each training dataset (Breiman 1996). In turn, the prediction results of all models are combined to address the final result. Majority voting is the selection method for this classification.

Algorithm 1 Multi-target Random Forest model training

Input:
$Tr \leftarrow$ training data set
$Ts \leftarrow$ testing data set
$RM \leftarrow$ random forests algorithm
Output:
$O \leftarrow$ a vector of predicted labels

```
1: Begin
2: for each t ← one target variable do
3:     RM.train(Tr.get(t))
4:     o ← RM.predict(Ts.get(t))
5:     O.add(o)
6: end for
7: return O
8: end
```

3.3 Clustering

A clustering approach is required to group the multi-target prediction results and recognize the driving patterns. For this, K-means (Wagstaff et al. 2001), is used to cluster the predicted results into k groups (i.e., driving pattern). K-means focuses on the similarity between variables (i.e., predicted labels) and uses a distance-based measurement approach to form cluster variables according to their nearest mean. For this, the Sum of Squared Errors = (SSE) is measured to find an optimal number of clusters. Equation 1 calculates SSE to find the best-fitted K, which minimizes the clustering cost, where x and m_i are, respectively, the value and cluster centre in cluster C_i. According to the results, K = 4 is the best number of clusters in this application.

$$SSE = \sum_{K=1}^{K} \sum_{x \in C_i} dist^2(m_i, x) \quad (1)$$

This study uses Eq. 2 to standardise data samples before the clustering. It results in increased clustering model accuracy by transforming data values into a mean of zero.

$$Rescaled\ Value = \frac{Value - min(Value)}{max(Value) - min(Value)} \quad (2)$$

Algorithm 2 uses K-means to find the similarities between the standardised target prediction results and group them into driving patterns. Figure 4 depicts the result of K-means as four clusters (driving pattern).

Algorithm 2 Driving pattern clustering using K-means

Input:
$Ts \leftarrow$ the standardize prediction results
$K \leftarrow 4$
$KM \leftarrow$ k-means algorithm
Output:
$C \leftarrow$ a vector of driving pattern clusters
1: **Begin**
2: **for each** $t \leftarrow$ the results of each target **do**
3: $c_t \leftarrow KM.predict(Ts.get(t))$
4: $C.add(c_t)$
5: **end for**
6: **return** C
7: **end**

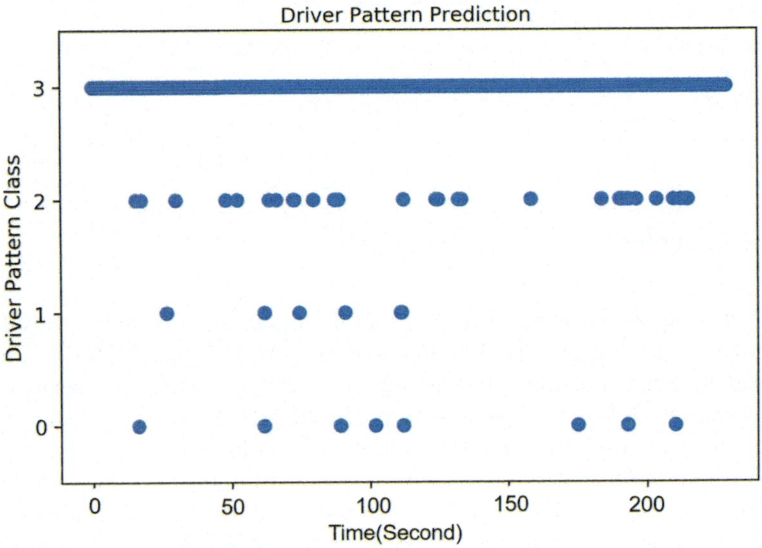

Fig. 4 Driving pattern clustering results

4 Result and Discussion

This section introduces the performance evaluation metrics and presents the results of the proposed model as compared to the benchmarks.

4.1 Performance Evaluation Metrics

There are four well-known performance evaluation metrics, including Accuracy, Precision, Recall, and F1 score, that are usually used to evaluate classification models (Davis and Goadrich 2006). They are introduced below:

- *Accuracy* is calculated by Eq. 3 to show the number of correct predictions according to all the predictions. For this, True Positive (TP), False Positive (FP), True Negative (TN), and False Negative (FN), are respectively true prediction as positive, false prediction as positive, true prediction as negative, and false prediction as negative.
$$Accuracy = \frac{TP + TN}{TP + TN + FN + FP}. \tag{3}$$
- *Precision* uses Eq. 4 to calculate the proportion of true positive samples over all positive predictions.
$$Precision = \frac{TP}{TP + FP} \tag{4}$$
- *Recall* measures the proportion of correctly predicted positives over all positive labels using Eq. 5.
$$Recall = \frac{TP}{TP + FN} \tag{5}$$
- *F1 score* uses Eq. 6 to take both precision and recall into account and study the performance of the prediction model.
$$F1\ Score = \frac{2 \times Precision \times Recall}{Precision + Recall} \tag{6}$$

4.2 The Model Evaluation

This section presents and compares the performance of the proposed model (i.e., the Random Forest) with four well-known classification models, including Decision Tree, Multilayer Perceptron, Multinomial Logistics Regression, and Logistics Regression-based one versus all classification.

The performance of the predictive models is evaluated in terms of Accuracy, Precision, Recall, and F1 score for each target. Table 2 presents the evaluation results based on the vehicle speed target. As the results show, the proposed Random Forest model outperforms the benchmarks. According to the correlation results, the Engine RPM, Shift number, and Cars Load are the most significant factors in this prediction.

Table 3 supports that the Random Forest model gives better prediction performance for rain intensity targets. As the correlation results show, Air Conditioning

Table 2 Performance evaluation (vehicle's speed)

Machine learning models	Accuracy	Precision	Recall	F1-score
The proposed Random Forest	0.842	0.831	0.842	0.820
Decision Tree	0.701	0.625	0.700	0.640
Multilayer Perceptron	0.608	0.588	0.608	0.536
Logistic Regression-based one versus rest classifier	0.566	0.554	0.567	0.557
Multinomial Logistics Regression	0.442	0.183	0.161	0.171

Table 3 Performance evaluation (rain intensity)

Machine learning models	Accuracy	Precision	Recall	F1-score
The proposed Random Forest	0.938	0.931	0.980	0.950
Decision Tree	0.642	0.684	0.533	0.599
Multilayer Perceptron	0.536	0.400	0.391	0.521
Logistic Regression-based one versus rest classifier	0.512	0.500	0.488	0.493
Multinomial Logistics Regression	0.423	0.346	0.412	0.376

Table 4 Performance evaluation (driver well-being)

Machine learning models	Accuracy	Precision	Recall	F1-score
The proposed Random Forest	0.992	0.995	0.996	0.996
Decision Tree	0.735	0.870	0.591	0.704
Multilayer Perceptron	0.602	0.600	0.484	0.536
Logistic Regression-based one versus rest classifier	0.533	0.466	0.437	0.451
Multinomial Logistics Regression	0.433	0.326	0.376	0.349

Status, Visibility, and Cars Load are the key variables that influence this target as compared to other predictors.

Table 4 evaluates the results of the proposed prediction model and shows that the Random Forest model outperforms the benchmarks for the driver's well-being prediction. Passenger Count, Cars Load, and Visibility are the key factors in this prediction.

The results of driver rush prediction are summarised in Table 5. According to it, the Random Forest model has better performance as compared to the benchmark. Driver rush prediction is highly correlated to Passenger Count, Cars Load, and Window Opening predictors based on correlation coefficient results.

Table 5 Performance evaluation (driver rush)

Machine learning models	Accuracy	Precision	Recall	F1-score
The proposed Random Forest	0.963	0.963	0.981	0.971
Decision Tree	0.843	0.886	0.526	0.660
Multilayer Perceptron	0.734	0.720	0.486	0.580
Logistic Regression-based one versus rest classifier	0.643	0.642	0.498	0.561
Multinomial Logistics Regression	0.539	0.478	0.451	0.464

Table 6 The average prediction delay (10,000 random selected samples)

Machine learning models	Model training delay (s)
The proposed Random Forest	15.41
Decision Tree	14.56
Multilayer Perceptron	16.22
Logistic Regression-based one versus rest classifier	15.87
Multinomial Logistics Regression	17.30

4.3 Model Training Delay

Model training delay is a metric measured since a model starts training until it calculates the outputs. According to it, better ML model efficiency can be achieved if the model training delay is minimized.

This study evaluates the average perdition delay of the proposed model. For this, a computing workstation with Intel® Xeon® Silver 4108 CPU (1.8 GHz per core), 128 G RAM, Windows 10 OS, and Jupyter Notebook (Python 3) is used to run the codes and train/test the models. It measures the average prediction delay of 10,000 random-selected samples that are repeated three times. As Table 6 summarizes, the proposed Random Forest model has better prediction delay as compared to the three benchmarks, including Multilayer Perceptron, Multinomial Logistics Regression, and Logistics Regression-based one versus all classification. However, it underperforms the Decision Tree model as the Random Forest forms and trains multiple trees instead of a single tree for prediction.

5 Conclusion and Future Work

This chapter proposes a global multi-target prediction model to recognize driving patterns. This factor is important in achieving or sustaining healthy and sustainable cities and communities. Applications are broad and could help propose context-specific solutions to resolve traffic-related issues. The model offers several benefits, such as proposing a global multi-target prediction model with a minimum training delay, building a data pre-processing for big transportation datasets, and evaluating the performance of the proposed model using an extensive data-driven approach. For this, a large-scale public dataset is used to train the machine learning models. This dataset is pre-processed using feature selection (i.e., Pearson's correlation coefficients), discretization (i.e., Quantile Discretize), and resampling (i.e., Hybrid) techniques to remove meaningless features and enhance data quality according to the Big data analytics requirements. A Random Forest technique is used to predict four targets, and then the KNN clustering approach groups the prediction vectors and forms the driving patterns. According to the results, the Random Forest technique outperforms the benchmark in terms of model training evaluation metrics (i.e., accuracy, precision, recall, and F1 score) and model training delay.

There are still a few issues that can be addressed in future work. First, the performance of this predictive analysis can be improved if the Federated Learning method is used to train the multi-target prediction model. It takes the benefits of distributed ML model training to enhance parallelisation and reduce model training delay. Local ML models are trained to predict each target in parallel, whereas the global model combines the results of slave models to predict multiple targets. Second, this research study can be extended if a larger number of driving features, such as driver's emotion and skill, are used for diving pattern classification. It allows for testing the impact of cognitive and emotional attributes on transportation services. Third, it is suggested to study the performance of other well-known multi-target ML prediction techniques (i.e., neural networks) in the field of transportation and driving pattern recognition.

References

Al Muktadir, Abu Hena, Takaya Miyazawa, Pedro Martinez-Julia, Hiroaki Harai, and Ved P. Kafle. 2019. Multi-target classification based automatic virtual resource allocation scheme. *IEICE Transactions on Information and Systems* 102 (5): 898–909.

Al-Ibrahim, Ali. 2011. Discretization of continuous attributes in supervised learning algorithms. *The Research Bulletin of Jordan ACM-ISWSA* 7952.

Andre, Michel. 1996. *Driving cycles development: Characterization of the methods*. SAE Technical Paper. Technical report.

Apache Spark. 2023a. https://spark.apache.org/. Retrieved March 2023.

Apache Spark. 2023b. Quantile discretizer. https://spark.apache.org/docs/latest/api/python/reference/api/pyspark.ml.feature.QuantileDiscretizer.html. Retrieved March 2023.

Ardakani, Saeid Pourroostaei, Tianqi Xia, Ali Cheshmehzangi, and Zhiang Zhang. 2022. An urban-level prediction of lockdown measures impact on the prevalence of the COVID-19 pandemic. *Genus* 78 (1). https://doi.org/10.1186/s41118-022-00174-6.

Ardakani, Saeid Pourroostaei, Xiangning Liang, Kal Tenna Mengistu, Richard Sugianto So, Xuhui Wei, Baojie He, and Ali Cheshmehzangi. 2023. Road car accident prediction using a machine-learning-enabled data analysis. *Sustainability* 15 (7): 5939. https://doi.org/10.3390/su15075939.

Bella, Francesco. 2014. Driver perception hypothesis: Driving simulator study. *Transportation Research Part F: Traffic Psychology and Behaviour* 24: 183–196.

Benesty, Jacob, Jingdong Chen, Yiteng Huang, and Israel Cohen. 2009. Pearson correlation coefficient. In *Noise reduction in speech processing*, 1–4. Springer.
Bernardi, Mario Luca, Marta Cimitile, Fabio Martinelli, and Francesco Mercaldo. 2018. Driver and path detection through time-series classification. *Journal of Advanced Transportation* 2018.
Breiman, Leo. 1996. Bagging predictors. *Machine Learning* 24 (2): 123–140.
Choi, Sang Jo, Jeong Hee Kim, Dong Gu Kwak, Pongtep Angkititrakul, and J.H. Hansen. 2007. Analysis and classification of driver behavior using in-vehicle can-bus information. In *Biennial Workshop on DSP for In-Vehicle and Mobile Systems*, 17–19.
Cutler, Adele, D. Richard Cutler, and John R. Stevens. 2012. Random forests. In *Ensemble machine learning*, 157–175. Springer.
Davis, Jesse, and Mark Goadrich. 2006. The relationship between precision-recall and ROC curves. In *Proceedings of the 23rd International Conference on Machine Learning*, 233–240.
Elgowainy, A., J. Han, L. Poch, M. Wang, A. Vyas, M. Mahalik, and A. Rousseau. 2010. Well-to-wheels analysis of energy use and greenhouse gas emissions of plug-in hybrid electric vehicles. https://doi.org/10.2172/982352.
Ericsson, Eva. 2000. Variability in urban driving patterns. *Transportation Research Part D: Transport and Environment* 5 (5): 337–354.
Ericsson, Eva. 2001. Independent driving pattern factors and their influence on fuel-use and exhaust emission factors. *Transportation Research Part D: Transport and Environment* 6 (5): 325–345. ISSN: 1361-9209.
Fernández, Alberto, Salvador Garcia, Francisco Herrera, and Nitesh V. Chawla. 2018. SMOTE for learning from imbalanced data: Progress and challenges, marking the 15-year anniversary. *Journal of Artificial Intelligence Research* 61: 863–905.
Higgs, Bryan, and Montasir Abbas. 2014. Segmentation and clustering of car-following behavior: Recognition of driving patterns. *IEEE Transactions on Intelligent Transportation Systems* 16 (1): 81–90.
Horowitz, Joel L. 2001. The bootstrap. In *Handbook of econometrics*, vol. 5, 3159–3228. Elsevier.
Kocev, Dragi, Celine Vens, Jan Struyf, and Sašo Džeroski. 2007. Ensembles of multi-objective decision trees. In *European Conference on Machine Learning*, 624–631. Springer.
Kocev, Dragi, Celine Vens, Jan Struyf, and Sašo Džeroski. 2013. Tree ensembles for predicting structured outputs. *Pattern Recognition* 46 (3): 817–833.
Kuhler, M., and D. Karstens. 1978. *Improved driving cycle for testing automotive exhaust emissions*. SAE Technical Paper. Technical report.
Last, Mark, Nitzan Rabinowitz, and Gideon Leonard. 2016. Predicting the maximum earthquake magnitude from seismic data in Israel and its neighboring countries. *PLoS ONE* 11 (1): e0146101.
Last, Mark, Sigal Elnekave, Amos Naor, and V. Shonfeld. 2008. Predicting wine quality from agricultural data with single-objective and multi-objective data mining algorithms. In *Recent advances on mining of enterprise data: Algorithms and applications*, 323–365.
Last, Mark. 2004. Multi-objective classification with info-fuzzy networks. In *European Conference on Machine Learning*, 239–249. Springer.
Levatić, Jurica, Dragi Kocev, Michelangelo Ceci, and Sašo Džeroski. 2018. Semi-supervised trees for multi-target regression. *Information Sciences* 450: 109–127.
Lin, Chan-Chiao, Soonil Jeon, Huei Peng, and Jang Moo Lee. 2004. Driving pattern recognition for control of hybrid electric trucks. *Vehicle System Dynamics* 42 (1–2): 41–58. https://doi.org/10.1080/00423110412331291553.
Malik, Meenakshi, Rainu Nandal, Surjeet Dalal, Vivek Jalglan, and Dac-Nhuong Le. 2021. Driving pattern profiling and classification using deep learning. *Intelligent Automation and Soft Computing* 28 (3): 887–906.
Neubauer, Jeremy, Aaron Brooker, and Eric Wood. 2013. Sensitivity of plug-in hybrid electric vehicle economics to drive patterns, electric range, energy management, and charge strategies. *Journal of Power Sources* 236: 357–364. ISSN: 0378-7753. https://doi.org/10.1016/j.jpowsour.2012.07.055.

Ramchoun, Hassan, Mohammed Amine Janati Idrissi, Youssef Ghanou, and Mohamed Ettaouil. 2016. Multilayer perceptron: Architecture optimization and training. *IJIMAI* 4 (1): 26–30.

Schneck, Arne, Sven Kalle, Rüediger Pryss, Winfried Schlee, Thomas Probst, Berthold Langguth, Michael Landgrebe, Manfred Reichert, and Myra Spiliopoulou. 2017. Studying the potential of multi-target classification to characterize combinations of classes with skewed distribution. In *2017 IEEE 30th International Symposium on Computer-Based Medical Systems (CBMS)*, 630–635. IEEE.

Song, Yan-Yan, and L.U. Ying. 2015. Decision tree methods: Applications for classification and prediction. *Shanghai Archives of Psychiatry* 27 (2): 130.

Spanos, Georgios, and Lefteris Angelis. 2018. A multi-target approach to estimate software vulnerability characteristics and severity scores. *Journal of Systems and Software* 146: 152–166.

Starkweather, Jon and Amanda Kay Moske. 2011. Multinomial logistic regression.

Tax, David M.J., and Robert P.W. Duin. 2002. Using two-class classifiers for multiclass classification. In *Object recognition supported by user interaction for service robots*, vol. 2, 124–127. IEEE.

Vitor, R.F. 2023. Car trips data log: An attempt to predict driver's behaviour. https://www.kaggle.com/datasets/vitorrf/cartripsdatamining. Retrieved March 2023.

Wagstaff, Kiri, Claire Cardie, Seth Rogers, Stefan Schroedl, et al. 2001. Constrained k-means clustering with background knowledge. *Icml* 1: 577–584.

Wang, Wenshuo, and Junqiang Xi. 2016. A rapid pattern-recognition method for driving styles using clustering-based support vector machines. In *2016 American Control Conference (ACC)*, 5270–5275. IEEE.

Wang, Wenshuo, Junqiang Xi, and Ding Zhao. 2018. Driving style analysis using primitive driving patterns with Bayesian nonparametric approaches. *IEEE Transactions on Intelligent Transportation Systems* 20 (8): 2986–2998.

Watson, Harry Charles. 1978. Vehicle driving patterns and measurement methods for energy and emissions assessment.

Yu, Hong-Zeng. 2012. Froude-modeling-based general scaling relationships for fire suppression by water sprays. *Fire Safety Journal* 47: 1–7. ISSN: 0379-7112. https://doi.org/10.1016/j.firesaf.2011.09.006.

Zhang, Jun, ZhongCheng Wu, Fang Li, Chengjun Xie, Tingting Ren, Jie Chen, and Liu Liu. 2019. A deep learning framework for driving behavior identification on in-vehicle CAN-BUS sensor data. *Sensors* 19 (6): 1356.

Zhou, Dengyong, Olivier Bousquet, Thomas N. Lal, Jason Weston, and Bernhard Schölkopf. 2004. Learning with local and global consistency. In *Advances in Neural Information Processing Systems*, 321–328.

Chapter 5
A Predictive Data Analysis for Traffic Accidents: Real-Time Data Use for Mobility Improvement and Accident Reduction

Abstract Road traffic accidents are a significant component of human and economic losses. In order to implement effective policies or carry out safe traffic travel and improve the road system's safety, it is essential to identify the data patterns in the dataset for extracting the main influencing features associated with traffic accidents. The analysis of traffic accidents is complex as they affect each other and are also affected by many other factors. We used the US Accidents dataset, covering 2.8 million accident records in 49 states in the United States from February 2016 to December 2021. Maximum Relevance (MR) is used to get feature relevance by obtaining Mutual Information (MI) of features. The traffic accident predictive models are expected to be more accurate by using feature relevance to assist people in making real-time transportation decisions to improve mobility and reduce accidents. The findings help improve urban transportation network and systems at multiple spatial levels.

Keywords Unsupervised learning · Traffic accident · Road system · Safety · Feature relevance · Mutual information

1 Introduction

Traffic congestion and accidents impact people's daily life in various ways. In all cases, road traffic accidents are a significant component of human and economic losses (Atubi and Gbadamosi 2015). According to the World Health Organisation (WHO), there were 1.35 million road traffic deaths globally in 2016, and between 20 and 50 million more people suffered non-fatal injuries or disabilities (World Health Organization 2023). Since the beginning of the 21st century, the United Nations (UN), the World Health Organisation (WHO), and the World Bank have all paid increasing attention to road safety issues (Peden and Prasanthi 2019). Such high-level attention indicates the necessity to make cities and communities healthier and safer, allowing opportunities for improved mobility and reduced car accidents.

The research work in this chapter was supported by our research team members: Zhihao Li, Yani Huang, Xi Wang.

S. Pourroostaei Ardakani and A. Cheshmehzangi, *Big Data Analytics for Smart Transport and Healthcare Systems*, Urban Sustainability,
https://doi.org/10.1007/978-981-99-6620-2_5

In order to implement effective policies or carry out safe traffic travel and improve the safety of the road system, it is vital to analyse the correlation between the causes of traffic accidents and the features that affect the occurrence of traffic accidents. The analysis of traffic accidents is complex because it is affected by many factors, such as road conditions, time of day/week, traffic intensity, weather conditions, and events that may change in real time (Wenqi et al. 2017). Environmental factors can also affect the occurrence of traffic accidents, such as temperature, rainfall, and other weather conditions/components. At the same time, humidity and wind speed also play a certain role (Joshua et al. 2010). Thus, we note the complexity is due to multiple aspects and factors that could affect accident occurrence.

In order to predict and analyse the correlation between accidents and traffic parameters such as weather conditions, regions, road types, speed, and time, Maximum Relevance (MR) is acquired via features' Mutual Information (MI) to determine feature correlations. This method helps to build accurate traffic accident predictive models to make real-time transportation decisions, improve mobility, and prevent accidents. In this regard, this chapter proposes an MR approach to analyse traffic accidents. In achieving this, a publicly-available US Accidents dataset (2016–2021) is used that covers 2.8 million accident records in 49 states of the United States from February 2016 to December 2021 (Sobhan et al. 2019). Out of all 45 identified variables, 12 of them are traffic attributes, 9 are location attributes, 11 are weather attributes, and 13 are point of interest (POI) attributes. In order to facilitate MR, a pre-processing approach is proposed to clean the dataset and convert the string values. In turn, features' MI scores are calculated to figure out the highly correlated features. This correlation calculation is analysed and evaluated using a Bayesian Network solution. This means the pre-processed dataset and the MR algorithm are used to analyse the intrinsic associations between multiple factors. Moreover, a Gaussian Parsimonious Bayesian model is proposed to assess the relationship between features with a high correlation to the target and the accuracy of the classifier.

The rest of this chapter is organised as follows. Section 2 reviews the literature, outlines the research gaps, and highlights the state-of-the-art solutions. Section 3 introduces the proposed approach and explains the experimental plan. Section 4 presents and discusses the results and key findings of this research study. Lastly, Sect. 5 concludes the chapter and highlights the issues that should be addressed in future research.

2 Literature Review

Global case study-based studies indicate the importance of big data use and analysis at multiple scales. For instance, using Ethiopian Road Traffic 18,288 accident datasets, Beshah and Hill (2010) utilises a dataset of 18,288 records to investigate the role of road-related vital factors. Decision trees, naive Bayes, and K-Nearest Neighbour classifiers are used to model accident severity. For each classifier, ten attributes were provided to Weka one by one, including the dependent class. The decision tree outperformed the Nave Bayes and K-NN classifiers' inaccuracy. Different policy-makers were given regulations based on 'Road Orientation,' 'Particular Area,' 'Road Separation,' 'Sub-city,' 'Road Junction,' and 'Accident Severity' features to reduce accident severity.

Saeid et al. (2023) aim to study road car accident data patterns and build and evaluate predictive machine learning models using decision trees, random forest, multinomial logistic regression, and naïve Bayes techniques. It uses a public dataset containing the UK's road accident data during 2005–2014, except for the year 2008. This research proposes a data pre-processing method that consists of data cleaning, feature extraction, data attribute generalisation, and data outlier removal. It highlights the impactful data features for road traffic prediction and reports the results of machine learning models to predict accident severity, the number of casualties, and the number of vehicles. According to the results, Random Forest (RF) gives the best results compared to other benchmarks.

In other studies, Juan et al. (2011) used a Bayesian Network to develop a model using data from the Spanish General Road Directorate on traffic accidents (DGT). The study concentrated on the province of Granada Highway. The records from rural highway 1536 were utilised to look for patterns in road traffic accidents. Variables that provide information on the cause and severity of an accident (such as, roadway information, weather information, accident information, driver data, injury severity, etc.) are taken into account. Because traffic accident variables are inherently interrelated, the study relied on Bayesian Networks (BN) to describe uncertain traffic accident factors.

Similarly, to uncover regular contributors' elements on road accident data in Tamilnadu city, Geetha et al. (2023) created a hybrid Decision Tree Artificial Neural Network. The study used data mining WEKA software tools to compare the new hybrid DTANN technique with a Decision tree (J48) and K-nearest Neighbour algorithms. DT and KNN were outperformed by the Hybrid Decision Tree Artificial Neural Network.

In other studies, Xi et al. (2016) used the analytic Hierarchy Process (AHP) to determine the critical components for road traffic accidents and the Apriori association technique to uncover the hidden correlation between road traffic accidents in the 2012 China dataset. Xi et al. (2016) believes using the Apriori method before detecting and deleting faulty factors may result in improper cause correlation, lowering the technique's accuracy. As a result, their study filters out several variables that have minor impact on traffic accidents.

In order to discover essential elements connected to road accidents, Kumar and Toshniwal (2015) compared Latent Class Clustering and K-mode clustering data mining algorithms. The number of clusters was determined using gap statistics using the AIC, BIC, and CAIC criterion. In addition, the researchers used FP Growth techniques on each cluster to check if there was any link between the variables in the road accident dataset. Their findings show that there was no significant difference in performance between LCC and K-Mode clustering on road accident data except for calculation speed. Both strategies reduced the dataset's variability, which was collected across the whole state of Uttarakhand. Because the link between accidents and factors impacting them can be established, regression analysis (such as linear regression models, negative binomial regression models, and Poisson regression models) has been the most popular technique in traffic accident investigation.

In addition, under the influence of factors such as vehicle type, traffic accidents, and traffic participants' type of statistical distribution, Jin-bao Zhao established urban road traffic accident analysis based on Bayesian network structure, combined with a network model and application joint tree inference engine in the accident site (Xi et al. 2016). Thus, specific approaches require the research above to investigate the leading elements impacting accidents and accident patterns. However, due to the unique meaning of traffic accident data, certain internal relationships between numerous factors are not easily quantified (Xi et al. 2016). Furthermore, as the database grows over time, the need for computation speed and accuracy in the algorithm becomes more apparent, emphasising the limitations of present models and algorithms. In light of such valuable studies, the paradigm proposed in this research study is based on cluster analysis with Maximum Relevance.

3 Methodology

3.1 Data Preprocessing

This research proposes a data pre-processing approach that aims to handle missing values, encode string attributes, and remove irrelevant attributes to prepare the dataset for further processing (Salvador et al. 2016). Figure 1 depicts the pre-processing workflow.

To simplify yet maintain the results' accuracy, we convert the standard time to floating-point numbers in seconds and replace the boolean values with 0 or 1. As for weather features, the original data set contains 128 different labels. In order to facilitate analysis and processing, we classified the 128 labels into ten categories. For example, 'Light Rain' and 'Light Freezing Drizzle' belong to the label 'Rain'. Features such as Precipitation that contains nearly 19% null values are dropped. Features such as country, state, turning loop, and source, which have a single unique value each, and those such as 'ID', and 'description' that have no impact on the prediction target are removed from the dataset (Joshi and Hadi 2015).

Null values take a negative impact on feature selection and machine learning model training (Cui et al. 2018). To deal with this, the proportions of nulls within each feature with 2,845,342 rows in total are calculated. As Table 1 shows, three features contain null values, and thus, they are removed. After deleting these three features, the null value proportions of the remaining features are less than 1% (Cui et al. 2018).

To avoid memory errors caused by NumPy to the large CSV file, the original dataset of $x_{TotalRow}$ rows is separated into n_{file} sub-files and handled separately using Eq. 1 where $x_{FileSize}$ is the size of each sub-file (Zhao et al. 2019). In this experiment, the sub-file size is set to 100,000 rows with 2,845,343 rows in total, and 29 sub-files are generated for the purpose of reducing memory space requirement.

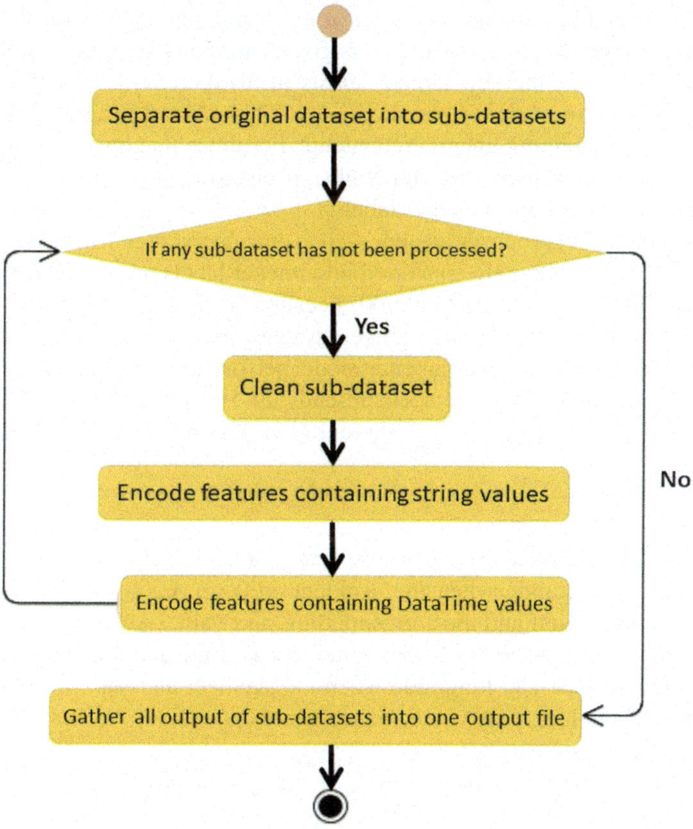

Fig. 1 Workflow for preprocessing

Table 1 Null value proportion for feature 'Number', 'Wind_Chill(F)' and 'Precipitation(in)'

Feature name	Null value proportion (%)
Number	61
Wind_Chill(F)	16
Precipitation(in)	19

$$n_{file} = \frac{x_{TotalRow}}{x_{FileSize}} \tag{1}$$

For features that contain meaningful string values, contents are encoded and converted into the corresponding numeric labels (Kim et al. 2020). This is implemented using a global dictionary list throughout the entire pre-processing procedure. Each feature has its own dictionary from the dictionary list (Abdulzahra et al. 2020). If strings do not exist in the dictionary, a new key-value pair will be added to the dictionary. The original string and the order of the string that appears for the first time in the form of an increasing integer label is given based on the previous integer label. Instead, if a string, as the key, already exists in the dictionary, it will be replaced with its corresponding value in the dictionary. This method is applied for features of string values containing uncertain contents. For those that have definite contents (e.g., only Boolean value or Day and Night), if-else statements are implemented to encode the string values into integer labels.

Zhang et al. (2019) proposes an approach to encode features with a DateTime format. The DateTime values are converted into integer labels presenting seconds from a fixed DateTime '2016-02-01 00:00:00' because this is earlier than any DateTime in the entire dataset. This method reduces storage space and loads the data into the model more efficiently. To reach a better model performance, the feature 'End_time', which means the time when the accident ends up, is replaced with the duration of each accident. As Eq. 2 shows, D is the duration, E is the 'End_time', and S is the 'Start_time'.

$$D_i = E_i - S_i \quad (i \in [1, \text{Total\_Row}]) \tag{2}$$

A similar method is applied to the features 'End_Lat' and 'End_Lng', which represent the longitude and latitude when the accident ends up. Each is replaced with the difference value of itself and the corresponding start value. According to Eqs. 3 and 4, D_lat and D_lng are the different values for latitude and longitude, E_lat and E_lng are the latitude and longitude when the accident ends up, S_lat and S_lng are the latitude and longitude when the accident happens.

$$D_{lat_i} = E_{lat_i} - S_{lat_i} \quad (i \in [1, \text{Total\_Row}]) \tag{3}$$

$$D_{lng_i} = E_{lng_i} - S_{lng_i} \quad (i \in [1, \text{Total\_Row}]) \tag{4}$$

Besides, other features containing meaningless strings that will hardly help in the subsequent steps are deleted: 'ID', 'Description', 'Street', 'City' (too distributed after encoded into integer labels), 'Zip_code', 'Country' (all are 'US') and 'Airport_code' (Joshi and Hadi 2015). Finally, we gather all output of sub-datasets into one output file for the next step.

3.2 *Information Correlation Calculation*

To extract the target in the US traffic accident dataset, feature selection methodology is investigated based on mutual information for both discrete and continuous data (Peng et al. 2005). **Definition Maximum Relevance (MR)** is used to get feature relevance by obtaining Mutual Information (MI) of features. Mutual information is a generic measure of the degree to which two variables are dependent on one another (Moon et al. 1995). The definition of mutual information $I(x; y)$ is shown as follows:

Definition [**Mutual Information** (Shannon 1948)]. Assume $p(x)$, $p(y)$ and $p(x, y)$ are the marginal and joint probabilistic density functions of two variables x and y. Their mutual information $I(x; y)$ is defined as Eq. 5.

$$I(x; y) = \iint p(x, y) \log \frac{p(x, y)}{p(x)p(y)} dxdy. \tag{5}$$

According to this, the result equals zero if two random variables are independent, while larger values indicate a greater degree of reliance (Guyon and Elisseeff 2003). Therefore, to choose the target which contains the maximum information about the rest of the features, the target selection can be reduced to maximise the joint dependency between features and the selected target. In an N-dimensional input space, target c can be calculated as Eq. 6.

$$c = \text{argmax}_{1 \leq k \leq N} \sum_{i \in \{1,\ldots,N\}} I(x_k; x_i). \tag{6}$$

The estimation of Mutual Information (MI) is different for continuous feature variables and discrete feature variables. Hence, Eq. 7 is used to calculate their MI when both variables are discrete (Vinh et al. 2010).

$$I(X; Y) = \sum_{y \in Y} \sum_{x \in X} p(x, y) \log \left(\frac{p(x, y)}{p(x)p(y)} \right). \tag{7}$$

In terms of continuous-continuous MI estimator, the original method is to partition the X and Y into bins of finite size using Eq. 8.

$$I(X, Y) \approx I_{\text{binned}}(X, Y) \equiv \sum_{ij} p(i, j) \log \frac{p(i, j)}{p_x(i)p_y(j)}. \tag{8}$$

A more sophisticated technique is to use a k-nearest neighbour entropy estimate instead of standard binning-based estimators (Kraskov et al. 2004). Finally, Brian (2014) extended the kth-neighbour method for MI between discrete and continuous variables based on the weighted Jensen-Shannon (JS) divergence estimate. By determining the kth-nearest neighbour to point i among the N_{x_i} data points whose discrete variable value matches x_i using a distance measure, the distance d between this kth neighbour and the origin is defined. According to Eq. 9, the number of neighbours m_i is counted in the entire data set who are within d of point i. With N_{x_i} and m_i,

$$\begin{aligned} I(X, Y) &= \langle I_i \rangle \\ &= \psi(N) - \langle \psi(N_{x_i}) \rangle + \psi(k) - \langle \psi(m_i) \rangle \end{aligned} \tag{9}$$

where $\psi(\cdot)$ is the digamma function (Abramowitz and Stegun 1965).

4 Result and Discussion

4.1 Data Analysis

After pre-processing the data, we plotted the processed data into graphs to visualise the relationship between different features and accidents. Next, we analyse the features in the dataset. In this regard, we note there are four levels of severity depending on the severity of the accident, where a smaller severity value indicates a shorter delay to nearby traffic and a larger value indicates a longer delay to traffic (Sobhan et al. 2019). As highlighted in Fig. 2, the vast majority of accidents were classified as level 2, with 2,382,472 accidents, representing 89.7% of the total. In order to make better use of the relationship between 'Start_time' and 'End_time', they are subtracted to obtain a new feature, Accident Duration, which plays an important role in how traffic accidents are handled, such as the speed of rescue, clearing traffic jams and reducing safety hazards (Xiaofeng 2011). Figure 3 shows the top 20 durations of the total number of accidents for the different durations. The duration of the highest number of incidents is 21,600 s or 6 h. Knowing the duration of most incidents, allows traffic management to more accurately allocate the time needed to respond and clear incidents to avoid unnecessary consumption of workforce and other resources (Nam and Mannering 2000).

Figure 4 shows the relationship between latitude and longitude and the number of accidents. The different sizes of the dots are used to differentiate between latitude and longitude and the number of accidents, with smaller dots indicating fewer accidents at the current latitude and longitude and larger dots indicating more accidents at the

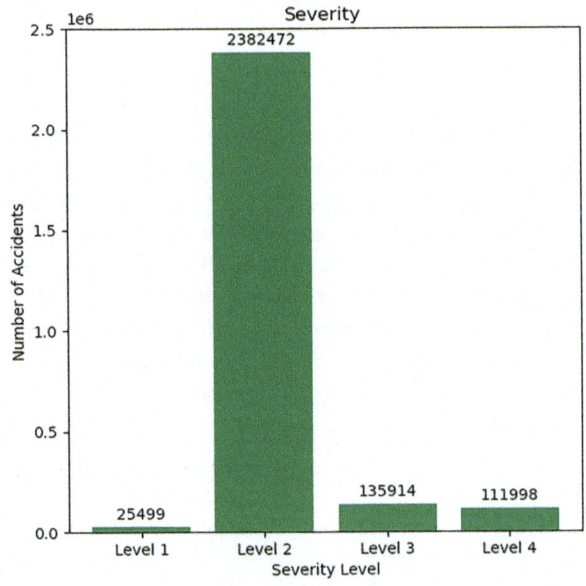

Fig. 2 Accidents severity

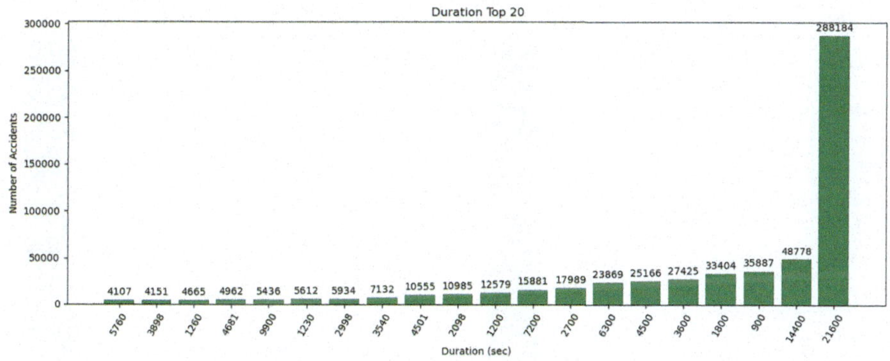

Fig. 3 Accidents duration top 20

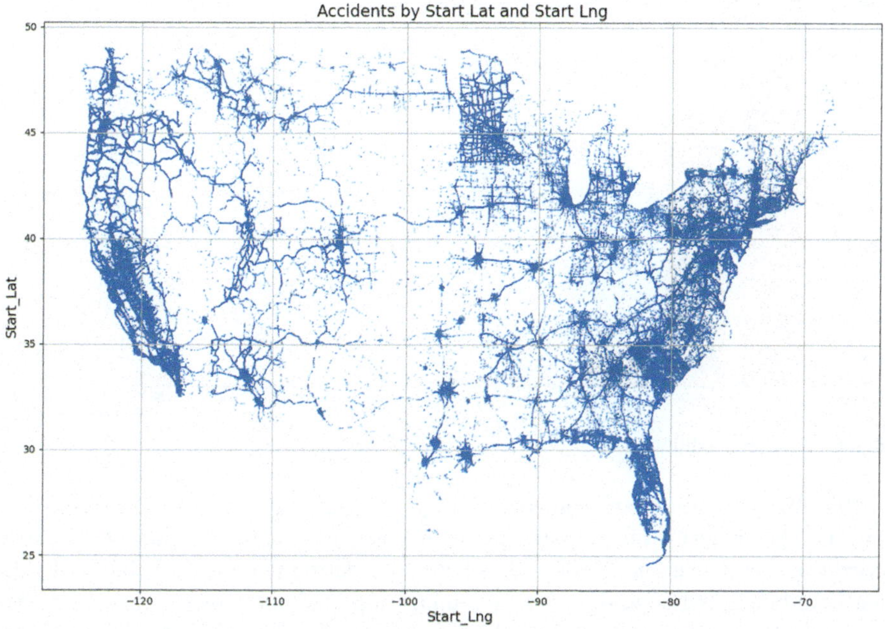

Fig. 4 Start latitude and start longitude

current point. The density of the dots also shows the approximate distribution of accident locations on the US map. Similar to Duration, the difference between the latitude and longitude at the start and the end is calculated and saved as 'Lat_Diff' and 'Lng_Diff', two features that visualise the distance travelled between the start and the end of the current incident. In Fig. 5, we found that in almost all accidents, the difference in latitude and longitude between the beginning and the end of the accident remained constant.

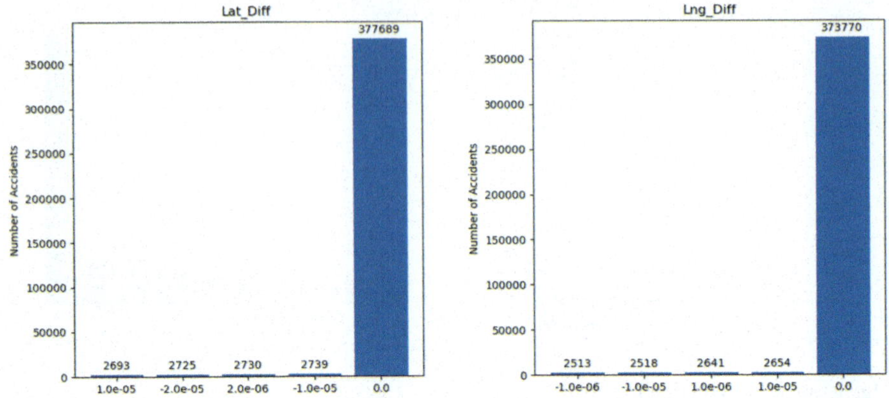

Fig. 5 Latitude difference and longitude difference

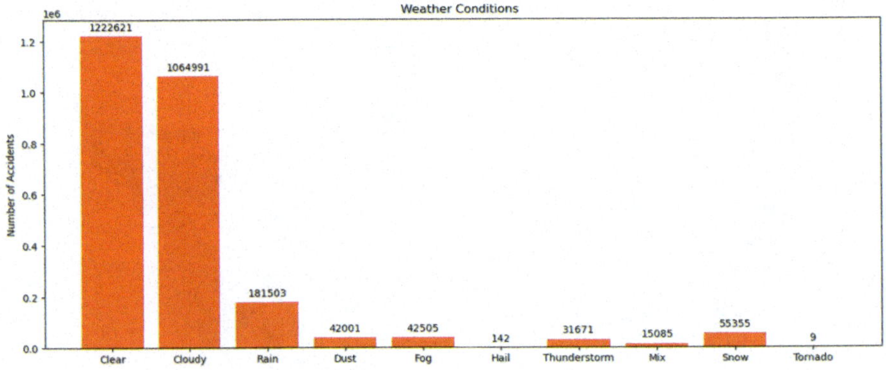

Fig. 6 Weather condition

Weather Condition was categorised in pre-processing to show the results as in Fig. 6. The majority of accidents are distributed in situations where the weather conditions are Clear and Cloudy. However, considering that the probability of clear and cloudy weather conditions is higher under normal conditions and that, according to the study, precipitation and temperature are also likely to be factors contributing to the occurrence of accidents, we further analysed the influence of other attributes such as humidity and temperature on the occurrence of accidents.

Figures 7, 8 and 9 focus on humidity, temperature, and pressure respectively. They depicts the top 15 accidents, and the corresponding scatters to show that the weather characteristics of humidity, temperature, and pressure have no significant impact on the occurrence of accidents. This is because the segments with a high probability of occurrence are located in the more routine (Scott et al. 2013), non-extreme weather range and have a more even distribution of accidents. In more extreme weather conditions, the number of accidents is lower. In Figs. 10 and 11, this is clear that there are more accidents with a wind speed of 0, a wind direction of 19, and a visibility of 10 mi.

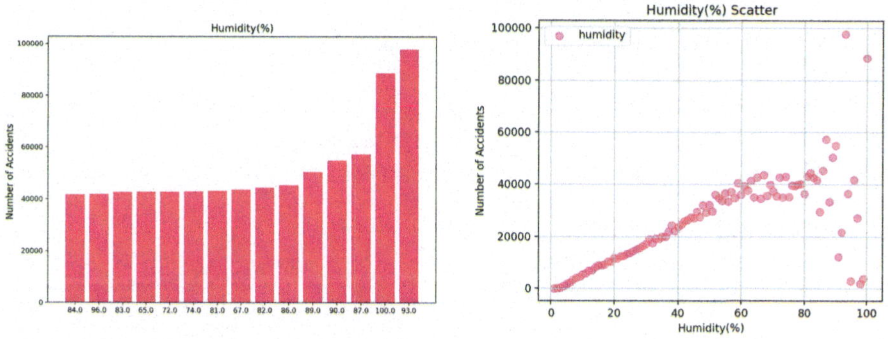

Fig. 7 Humidity histogram and scatter

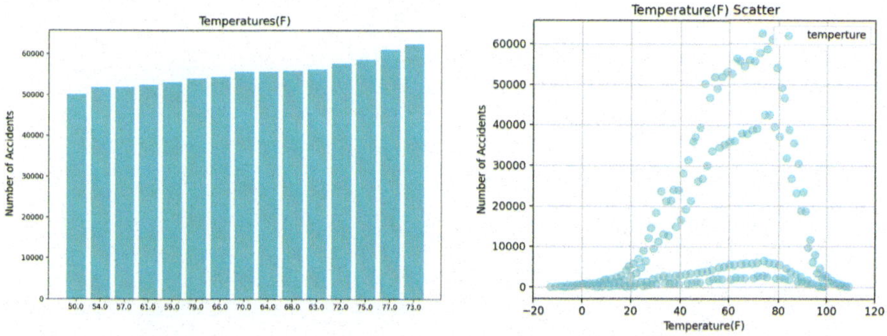

Fig. 8 Temperature histogram and scatter

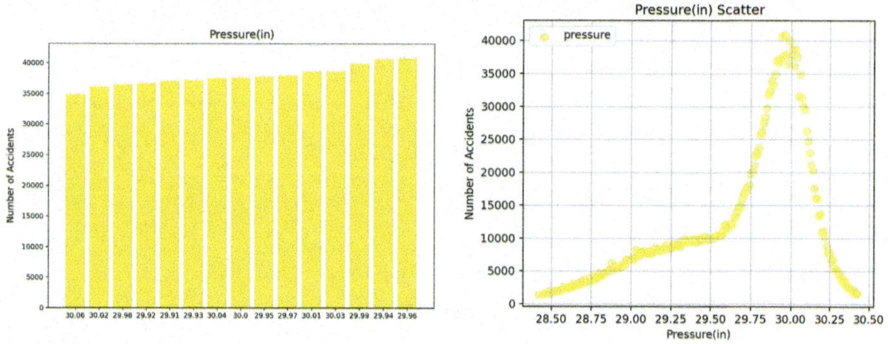

Fig. 9 Pressure histogram and scatter

However, other studies in the literature have shown that the higher the visibility, the lower the accident rate. This is because there is also a direct relationship between accident population density and accident rates, with densely populated areas often located in areas of high visibility such as cities (Janoff et al. 1978). In order to obtain

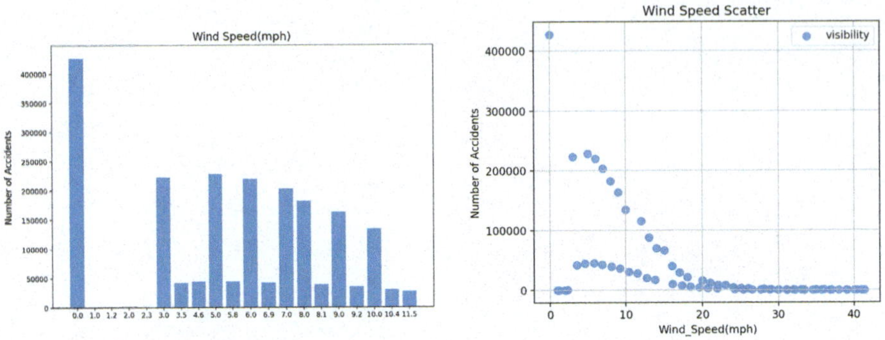

Fig. 10 Wind speed histogram and scatter

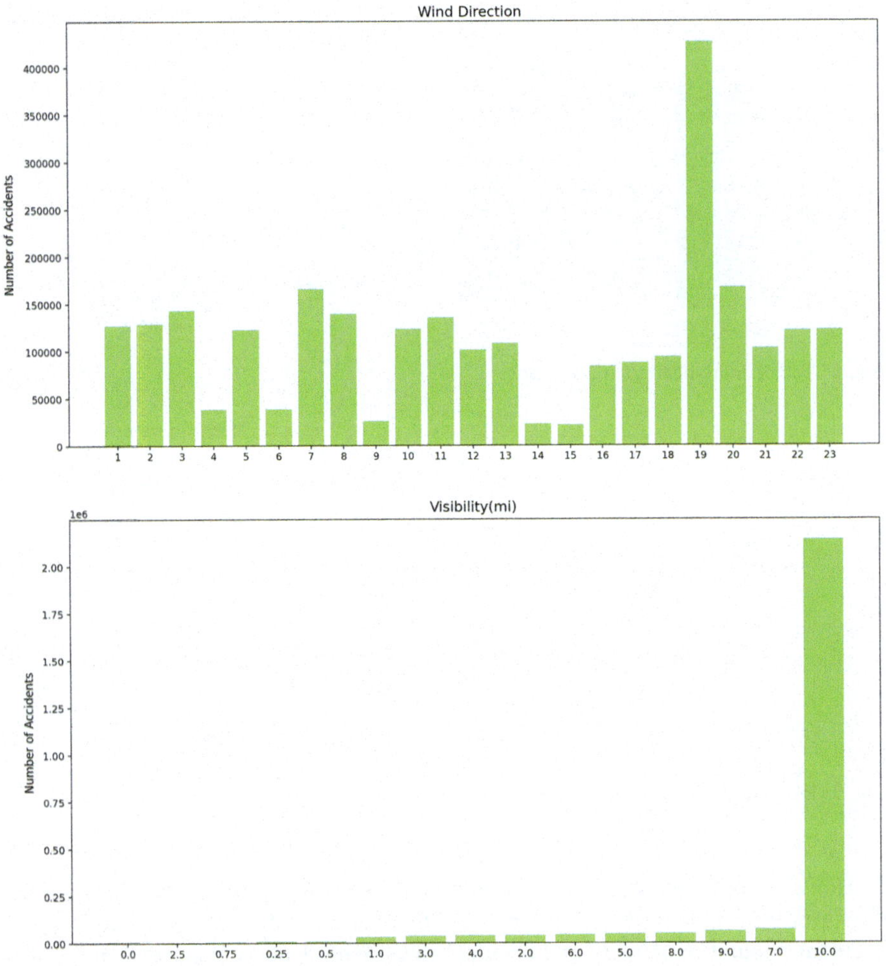

Fig. 11 Wind direction and visibility histogram

a more accurate basis for the analysis, in the following section, we will go further and use algorithms to analyse the relationship between accident features.

4.2 Correlation Calculation

Two main experiments are implemented. One is the total correlation for all features, and the other is for all features with the 'Severity' feature. The input data is the uniform sampling of 20,000 rows from the entire preprocessed dataset. Figures 12 and 13 are the results of these experiments with the metric of MI.

4.3 Evaluation

In order to verify the results, the Gaussian Naive Bayes model (Jahromi and Taheri 2017) is chosen to evaluate how the features with higher relevance to the target lead to higher accuracy of the classifier.

To make the comparison more reasonable and straightforward, two groups of features, both of which contain four bi-categorical features, are extracted from the whole dataset as inputs, and severity is selected as the target. The group called $good\_feature$ contains 'Sunrise_Sunset', 'Civil_Twilight', 'Junction' and 'Side'. The group called $bad\_feature$ contains 'Amenity', 'Bump', 'Crossing' and 'Railway'. Their mutual information scores with respect to severity are shown in Fig. 14.

According to the training scores and testing scores shown in Table 2, the features with higher mutual information scores provide better prediction results.

5 Conclusion

In the last decade or so, traffic accident analysis and prediction have become a popular research topic in multiple sub-fields of computer science. The findings are also important for urban-related research, particularly for traffic management, accident monitory checks, and other similar work. It is identified that a large amount of research focuses on the model to predict the target of traffic accidents, while only a few analyse the feature relationship within traffic accident datasets.

In this study, we filtered the original dataset of millions of entries to clean the missing values and encoded the meaningful string values into integer values. We used pre-processed dataset and MR algorithm to analyse the internal relationships between numerous factors. Throughout the experiment, methods like separating the dataset into sub-datasets were implemented to meet the requirement for computation speed and accuracy. Thus, we see the relevance of such findings for further optimisation of urban systems, urban networks, and transportation-related measures. For instance,

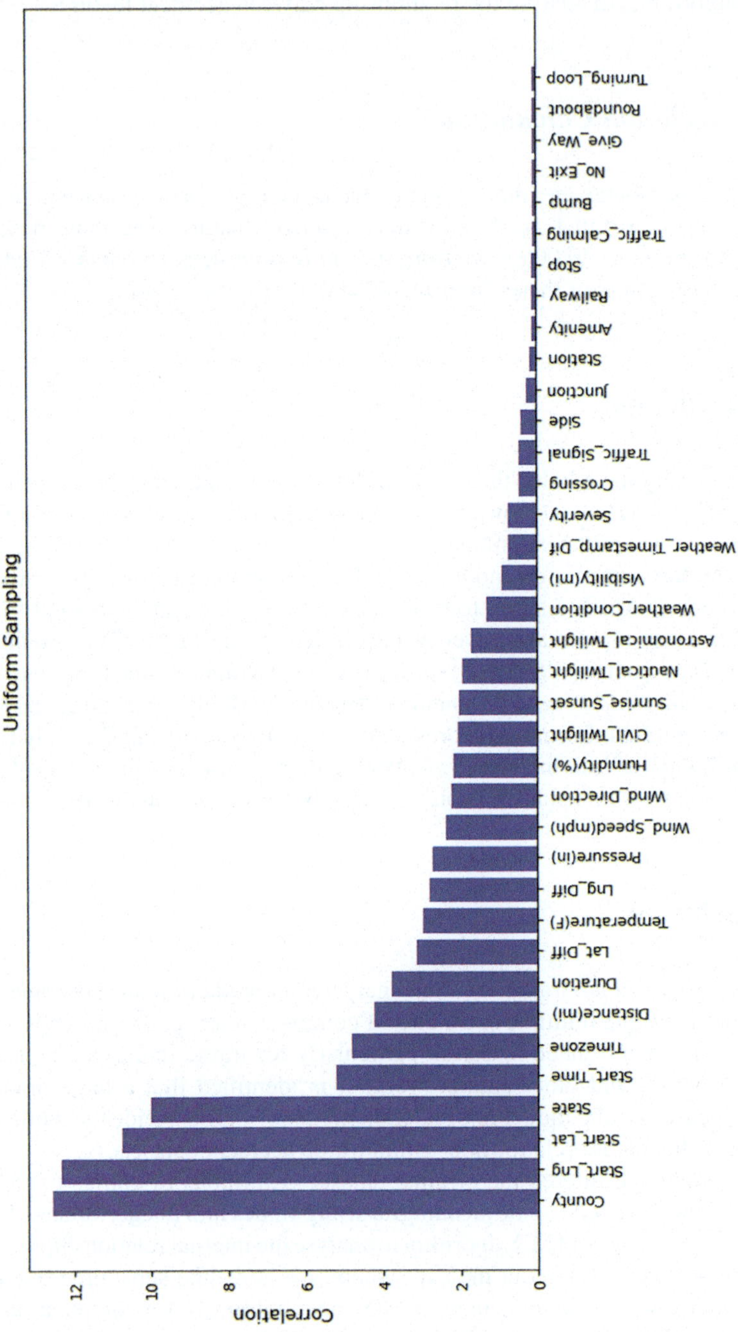

Fig. 12 Total mutual information

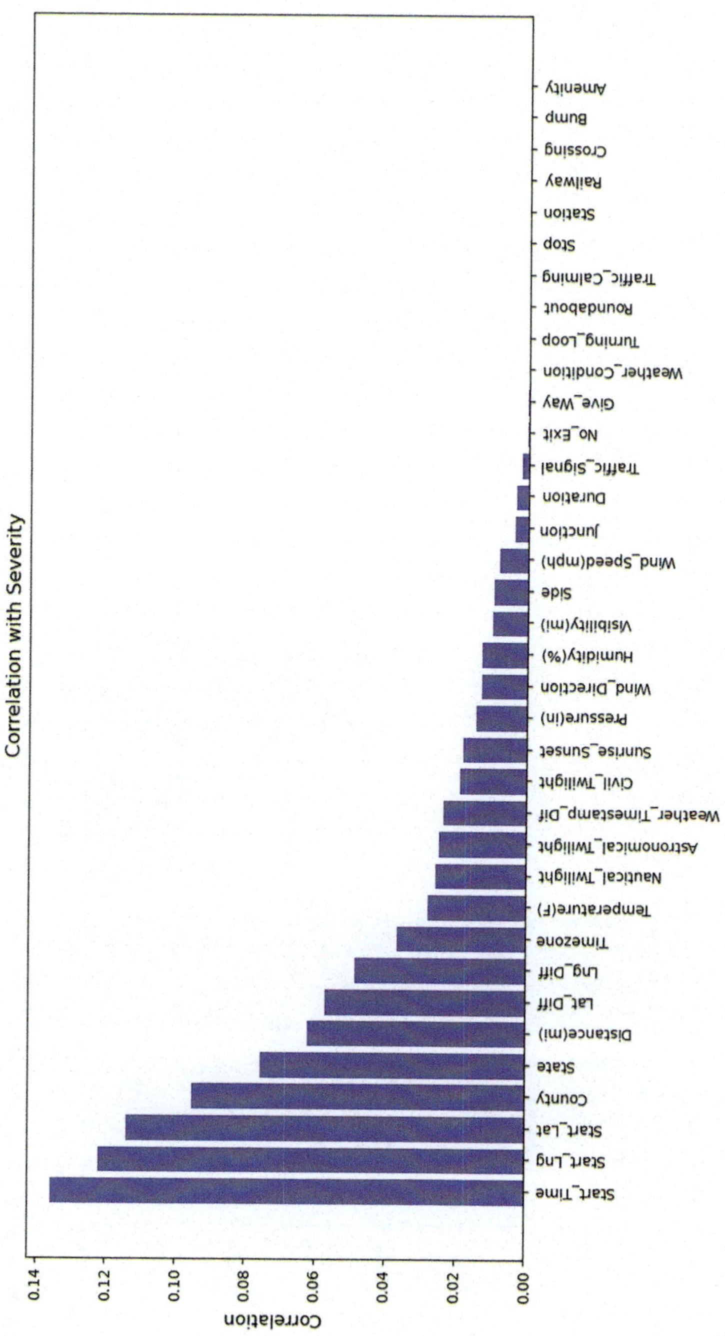

Fig. 13 Mutual information with severity

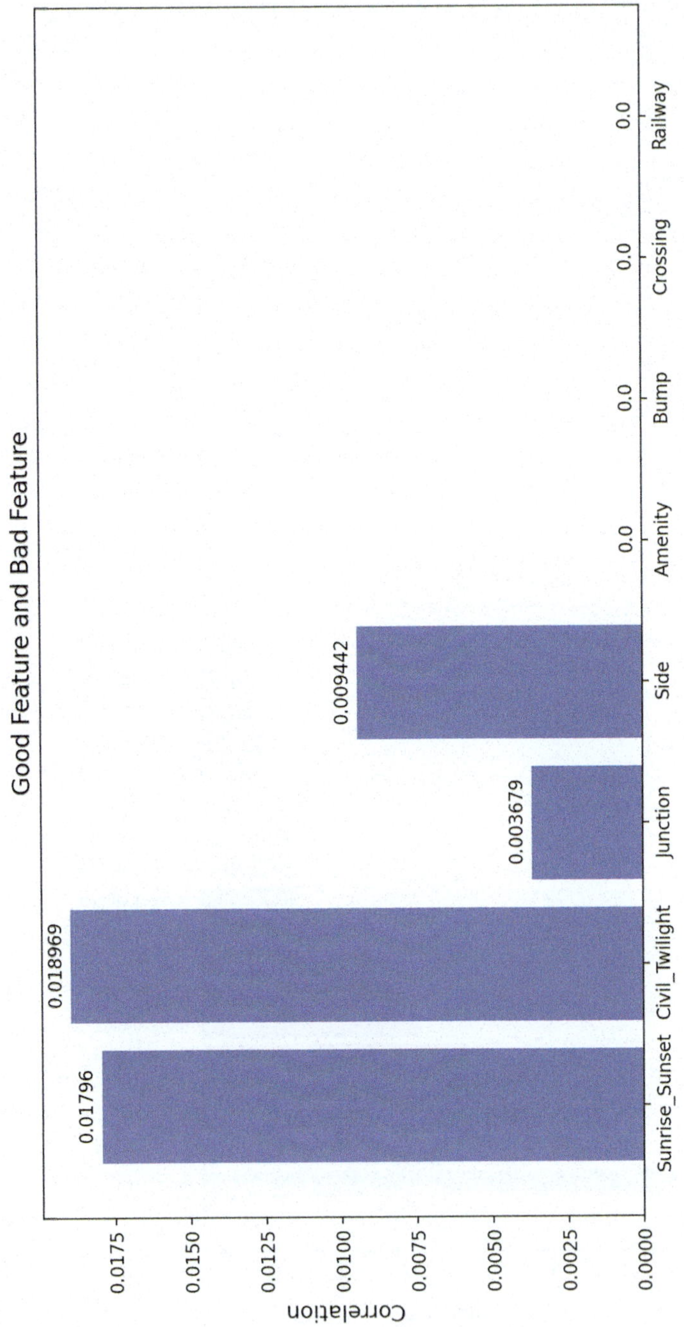

Fig. 14 MI score of selected features

Table 2 Predictive result comparison

Group	Train score	Test score
bad_feature	0.2601	0.2656
good_feature	0.3389	0.3382

the focus could be more on improving mobility patterns and behaviours as well as reducing the number of car accidents.

Lastly, we argue that the pre-processing methods used in this study are important for larger-scale studies, particularly when using big data. The study also indicates that the algorithms implemented and metrics used as the methods help evaluate the result of this experiment. Consequently, they help prove that our proposed model works in this research study. More importantly, the findings provide insight into present models and algorithms' feature relationships and potential optimisation approaches. These will be extremely important for future research focused on urban transportation networks and systems, as well as car accident reductions and mobility improvements.

References

Abdulzahra, Suha Abdulhussein, Ali Kadhum M. Al-Qurabat, and Ali Kadhum Idrees. 2020. Data reduction based on compression technique for big data in iot. In *2020 international conference on emerging smart computing and informatics (ESCI)*, 103–108. IEEE.

Abramowitz, Milton, and Irena A. Stegun. 1965. *Handbook of mathematical functions*. New York: Dover Publications.

Ardakani, Saeid Pourroostaei, Xiangning Liang, Kal Tenna Mengistu, Richard Sugianto So, Xuhui Wei, Baojie He, and Ali Cheshmehzangi. 2023. Road car accident prediction using a machine-learning-enabled data analysis. *Sustainability* 15 (7): 5939. https://doi.org/10.3390/su15075939.

Atubi, A.O., and K.T. Gbadamosi. 2015. Global positioning and socio-economic impact of road traffic accidents in Nigeria: Matters arising. *American International Journal of Contemporary Research* 5 (5): 136–146.

Aworemi, Joshua Remi, Ibraheem Adegoke Abdul-Azeez, and Segun Oluwaseun Olabode. 2010. Analytical study of the causal factors of road traffic crashes in southwestern Nigeria. *Educational Research* 1 (4): 118–124.

Beshah, Tibebe, and Shawndra Hill. 2010. Mining road traffic accident data to improve safety: Role of road-related factors on accident severity in Ethiopia. In *2010 AAAI spring symposium series*.

Cui, Zhiyong, Ruimin Ke, Ziyuan Pu, and Yinhai Wang. 2018. Deep bidirectional and unidirectional LSTM recurrent neural network for network-wide traffic speed prediction. arXiv preprint arXiv:1801.02143.

De Oña, Juan, Randa Oqab Mujalli, and Francisco J. Calvo. 2011. Analysis of traffic accident injury severity on Spanish rural highways using Bayesian networks. *Accident Analysis and Prevention* 43 (1): 402–411.

García, Salvador, Sergio Ramírez-Gallego, Julián Luengo, José Manuel Benítez, and Francisco Herrera. 2016. Big data preprocessing: Methods and prospects. *Big Data Analytics* 1 (1): 1–22.

Geetha, K., Statistical Inspector, and Rabiya Basri. 2023. Road accidents and road safety measures in Tamil Nadu: An analysis. https://tnsta.gov.in/tnsta/pdf/ra5.pdf. Retrieved February 2023.

Guyon, Isabelle, and André Elisseeff. 2003. An introduction to variable and feature selection. *Journal of Machine Learning Research* 3 (Mar): 1157–1182.

Jahromi, Ali Haghpanah, and Mohammad Taheri. 2017. A non-parametric mixture of Gaussian Naive Bayes classifiers based on local independent features. In *2017 artificial intelligence and signal processing conference (AISP)*, 209–212. IEEE.

Janoff, M.S., B. Koth, W. McCunney, M.J. Berkovitz, and M. Freedman. 1978. The relationship between visibility and traffic accidents. *Journal of the Illuminating Engineering Society* 7 (2): 95–104.

Joshi, Manish, and Theyazn Hassn Hadi. 2015. A review of network traffic analysis and prediction techniques. arXiv preprint arXiv:1507.05722.

Kim, Aechan, Mohyun Park, and Dong Hoon Lee. 2020. Ai-ids: Application of deep learning to real-time web intrusion detection. *IEEE Access* 8: 70245–70261.

Kraskov, Alexander, Harald Stögbauer, and Peter Grassberger. 2004. Estimating mutual information. *Physical Review E* 69 (6): 066138.

Kumar, Sachin, and Durga Toshniwal. 2015. A data mining framework to analyze road accident data. *Journal of Big Data* 2 (1): 1–18.

Moon, Young-Il, Balaji Rajagopalan, and Upmanu Lall. 1995. Estimation of mutual information using kernel density estimators. *Physical Review E* 52 (3): 2318.

Moosavi Sobhan, Mohammad Hossein Samavatian, Srinivasan Parthasarathy, Radu Teodorescu, and Rajiv Ramnath. 2019. Accident risk prediction based on heterogeneous sparse data: New dataset and insights. In *Proceedings of the 27th ACM SIGSPATIAL international conference on advances in geographic information systems*, 33–42.

Nam, D., and F. Mannering. 2000. An exploratory hazard-based analysis of highway incident duration. *Transportation Research Part A Policy and Practice* 34 (2): 85–102.

Peden, Margaret M., and Prasanthi Puvanachandra. 2019. Looking back on 10 years of global road safety. *International Health* 11 (5): 327–330.

Peng, Hanchuan, Fuhui Long, and Chris Ding. 2005. Feature selection based on mutual information criteria of max-dependency, max-relevance, and min-redundancy. *IEEE Transactions on Pattern Analysis and Machine Intelligence* 27 (8): 1226–1238.

Ross, Brian C. 2014. Mutual information between discrete and continuous data sets. *PLoS One* 9 (2): e87357.

Shannon, Claude Elwood. 1948. A mathematical theory of communication. *The Bell System Technical Journal* 27 (3): 379–423.

Stevens Scott E., Liqiang Sun, Emily Janssen, Donald Wuebbles, Michael C. Kruk, Devin Thomas, Martha Shulski, Natalie A. Umphlett, Kenneth G. Hubbard, Kevin Robbins, Luigi Romolo, Adnan Akyuz, Tapan B. Pathak, Tony R. Bergantino, Kenneth E. Kunkel, Laura E. Stevens, and J. Greg Dobson. 2013. Regional climate trends and scenarios for the US national climate assessment part 4. Climate of the US great plains. *NOAA Technical Report NESDIS 142-4*.

Vinh, Nguyen Xuan, Julien Epps, and James Bailey. 2010. Information theoretic measures for clusterings comparison: Variants, properties, normalization and correction for chance. *The Journal of Machine Learning Research* 11: 2837–2854.

Wenqi Lu, Luo Dongyu, and Yan Menghua. 2017. A model of traffic accident prediction based on convolutional neural network. In *2017 2nd IEEE international conference on intelligent transportation engineering (ICITE)*, 198–202. IEEE.

World Health Organization. 2023. Global status report on road safety 2018: Summary. https://www.who.int/publications/i/item/9789241565684. Retrieved February 2023.

Xi, Jianfeng, Zhonghao Zhao, Wei Li, and Quan Wang. 2016. A traffic accident causation analysis method based on Ahp-apriori. *Procedia Engineering* 137: 680–687.

Xiaofeng, W.U. 2011. Prediction model of freeway traffic accidents duration and severity. In *China public security*, Academy Edition.

Zhang, Chuanting, Haixia Zhang, Jingping Qiao, Dongfeng Yuan, and Minggao Zhang. 2019. Deep transfer learning for intelligent cellular traffic prediction based on cross-domain big data. *IEEE Journal on Selected Areas in Communications* 37 (6): 1389–1401.

Zhao, Lu, Yonghua Zhou, Huapu Lu, and Hamido Fujita. 2019. Parallel computing method of deep belief networks and its application to traffic flow prediction. *Knowledge-Based Systems* 163: 972–987.

Part II
Smart Healthcare

Chapter 6
Healthcare Infrastructure Development and Pandemic Prevention: An Optimal Model for Healthcare Investment Using Big Data

Abstract This chapter proposes an optimal model for predicting the minimum healthcare investment required for pandemic prevention. It focuses on the COVID-19 pandemic as a recent example to further analyse the role of healthcare infrastructure development. It highlights the relationship between the control of COVID-19 and infrastructure development using linear regression, SVR, KNN, and decision tree models. By analysing the impact of each feature on the results, we select appropriate attributes and data processing methods. The findings of this study demonstrate that KNN is the best model with the highest training score of 0.655, while the training result of the decision tree is the worst, and the score is no more than 0.5. Lastly, the study highlights how big data could be used to improve the availability and development of urban critical infrastructures, such as healthcare infrastructure. An optimal model is suggested as part of the conclusion of this study.

Keywords Healthcare · Infrastructure development · COVID-19 · Big data analysis · Machine learning

1 Introduction

The COVID-19 pandemic caused an unprecedented global crisis for more than three years. The situation put millions at risk of death, while the control of the pandemic by the medical systems of various countries differed from context to context. It is also evident that we will witness future outbreaks/pandemics and other disease variants. Thus, the application of big data is critical for detailed analysis to reveal the intrinsic links between virus control and other factors (Cheshmehzangi et al. 2023), such as critical infrastructure development like healthcare.

Existing research has focused on long-standing disparities and inequalities in public health spending across countries. Such studies sought to understand how health systems can be strengthened by evaluating their capacity to control COVID-19 (Lal et al. 2021) and other disruptive disease outbreaks. Some studies go further,

The research work in this chapter was supported by our research team members: Xuze LI, Longwen HU, Xiang GAO, and Jing WANG.

S. Pourroostaei Ardakani and A. Cheshmehzangi, *Big Data Analytics for Smart Transport and Healthcare Systems*, Urban Sustainability,
https://doi.org/10.1007/978-981-99-6620-2_6

seeking to explore the impact of governance structures, healthcare infrastructure, and lessons learned from previous pandemics on a country's response to the COVID-19 pandemic (Sharma et al. 2021). However, academic research in this field also has certain limitations. For example, they pay more attention to the relationships between the national governance structure for pandemic control and policy-related matters (Ardakani et al. 2022). Hence, in most cases, the literature regards the impact of healthcare infrastructure development as a less influential factor. Besides, the tracking data of COVID-19 signified dynamic changes, and the analysis and prediction made in the early stage of the epidemic are no longer time-sensitive.

To address these research gaps, we mainly focus on exploring the relationship between global healthcare infrastructure and COVID-19 pandemic control. Based on the findings of Sharma et al. (2021), we also propose a conceptual framework to find the links between public healthcare expenditure, healthcare infrastructure, and pandemic control. In this framework, health infrastructure is defined as the number of health workers and specialist surgical workforce, whereas health expenditure is represented by public health expenditure expressed in various proportions. In this study, we consider the case fatality around the world as an indicator of pandemic control. For this, annual data from the World Bank is collected to describe countries' healthcare infrastructure investments (World Bank 2023), and the WHO's COVID-19 global data is used to assess countries' ability to manage the outbreak (World Health Organization 2022; John Hopkins University 2022). Through secondary data collected as of April 25, 2022, and using a series of methods such as linear regression, SVM, and KNN, we analyse the correlation between the local epidemic control and the healthcare infrastructure investments of each country around the world. In doing so, big data analysis makes relevant predictions for the prevention and control of similar pandemics or outbreaks in the future.

The contribution of this chapter is mainly to study the emerging field, finding the correlation between healthcare infrastructure investments and pandemic control. It highlights the importance of critical infrastructures such as healthcare; thus, expanding investment in such infrastructure is crucial for future outbreak control or even prevention. This study uses and compares the pros and cons of multiple regression models, which help develop a set of guidelines for future pandemic control of similar scenarios.

This chapter is organised as follows. Section 2 reviews the literature, Sect. 3 describes the research methodology, including data pre-processing, and model training, Sect. 4 discusses the results, and Sect. 5 concludes the critical points of this research and highlights the issues that should be addressed as further work.

2 Literature Review

The COVID-19 pandemic placed enormous strain on countries around the world, exposing the huge differences in public health infrastructure and pandemic control. We highlight a few critical questions that need to be highlighted as part of this study:

1. What are the factors of pandemic prevention?
2. What can be considered a healthcare investment?
3. How healthcare investment affects pandemic prevention?
4. What is the relationship between healthcare investment on pandemic prevention?
5. What are the attributes in healthcare prevention contributing to pandemic prevention?
6. How much does each attribute weigh to affect pandemic prevention?

Leung et al. (2020) claimed that public health expenditure has a positive effect on the number of diagnosed and critical cases associated with pandemics like the COVID-19. However, it has no effect on the number of active or death cases. This fact makes it difficult to analyse the situation of the COVID-19 in any framework. For example, since one of the early pandemics of the 21st century, i.e., SARS, the global situation has changed dramatically. Therefore, finding the appropriate data and criteria to evaluate countries' public health systems is critical to studies like this. According to the article Lal et al. (2021), different countries have different policy tendencies in building their own public health systems. Depending on their specific national circumstances, health systems tend to have more extensive investments in global health security (GHS) capacities or universal health coverage (UHC) capacities. In the GHS propensity systems, public health spending is mainly used for preventing, detecting, and responding to public health threats, while the UHC propensity systems focus more on covering primary healthcare. It should be noted that some countries focus on strengthening one aspect while ignoring the other.

African countries, such as Nigeria, suggest an over-concentration of GHS efforts in response to tropical infectious diseases but still remain low in primary health care. Thus, such countries experienced ineffective control of COVID-19 due to a lack of intensive care unit beds and ventilators. In this regard, different emphases should be reflected in the selection of data, which is also highlighted in this study. Another exploratory study Diop and Asongu (2020) assesses differences in healthcare infrastructure in Africa and the world through big data analysis. This study lists some indicators that are very meaningful to our study here. The first dataset is from the United Nations Sustainable Development Goals (UN SDGs), including the selected number of people for every single doctor (Medical Doctor), the number of people for every single nurse or midwife (Nurse/Midwife), International Health Regulation scores (IHR), Healthy Life Expectancy (HALE), Universal Health Coverage (UHC), and Domestic General Government Health Expenditure (GGHE). The other dataset is related to the COVID-19 pandemic, including the number of Hospital beds (Hosp_bed), the number of Intensive Care Unit beds (ICU_bed), and the number of ventilators. These indicators reflect the healthcare infrastructure development and the COVID-19 prevention in each country.

In order to demonstrate the minimum investment requirement on healthcare infrastructure to prevent pandemics, the relationship of healthcare infrastructure ad pandemic control needs to be clear. Sharma et al. (2021) claimed that pandemic prevention is affected by more factors, such as the government control policy, how they learn from the previous pandemic, and how they lead the public. In addition to the macro-level factors, they proposed models from both the proactive side and reactive side to conclude that the pandemic prevention is also influenced by government policy and the existence of the previous pandemic in the country. Moreover, their model also considered the development of healthcare infrastructure as one of the contributing factors and concluded that the increasing healthcare infrastructure (i.e., as a percentage of GDP) is essential in fighting against the pandemic or disease outbreaks. They benchmarked the investment in healthcare by considering the healthcare expenditure as a percentage of GDP and the number of physicians per 1000 population (Donald et al. 2012). Their study indicates that healthcare investment is crucial to pandemic prevention. However, they only concluded with the suggestion to improve the GDP spent on healthcare infrastructure instead of the details on what contributes to the healthcare investment. In light of such factors, this study aims to find the correlation of each attribute with pandemic prevention.

The prevention performance for different countries was based on case-fatality rate, test positivity rate, active cases, and an estimate of the detection rate. Kuster and Overgaard (2020) proposed a single, comprehensive metric, named the COVID-19 Testing Index, scaled from 0 to 100 in order to make inter-country comparison easier. In addition, Aleksandar et al. (2020) proposed the SmartResilience project, which adopts the concept of infrastructure elasticity as the ability to understand and predict the risks threatening the key functions of infrastructure, prepare for expected or unexpected destructive events, absorb/bear their impact in the best way, respond and recover from them, and adjust/transform the infrastructure or its operation according to the lessons learned, to reduce the vulnerability of critical infrastructure. In the SmartResilience method, indicators can be of any type, and they are converted into equal scales, that is, scores on a scale from 0 to 5. The index scores are summarised upward in the six-level hierarchical model, ending with the overall elasticity level of the key infrastructure or country. If the evaluation is repeated, the trend curve is obtained. If necessary, problems can be classified in various ways. In terms of calculation, this can be solved by weighting, and one or more indicators are used to measure each problem. This approach assumes that the resilience of the healthcare system plays a key role in assessing the overall resilience of other relevant critical infrastructures. One of the advantages of this approach is that each measure taken before and during COVID-19 can be described as an indicator in the elasticity matrix and become more transparent.

Common statistical methods used for predicting Influenza diseases include the 'generalised linear model', 'autoregressive comprehensive moving average', and 'generalised autoregressive moving average' (Marathe and Vullikanti 2013). However, they largely depend on the availability of training data. In addition, because they are purely data-driven, they do not capture potential causal mechanisms. Therefore, epidemiological dynamics affected by behavioural adaptation are often challenging to capture (Wang et al. 2020). From the literature, it is noted that the artificial neural network (ANN) is becoming more and more prominent in epidemic prediction. This is mainly because of its self-learning ability without prior knowledge (Adiga et al. 2020).

There are several models suitable for predicting and analysing COVID-19. The first one is data-driven models, which use statistical and machine learning methods to predict outcomes, such as the number of cases, mortality, and hospital demand. The most common mechanism model of disease transmission in the population is the SIR model (Adiga et al. 2020). The SIR model divides the population of N drugs into three groups. Each group corresponds to a disease state, which is: susceptibility (S), infection (I), and removal or recovery (R). Then, specific models specify how susceptible individuals become infectious and then recover. Second, Roseline and his team adopted the ordinary least squares estimator to measure the impact of travelling history and the spread of COVID-19 in Nigeria and made a prediction about the correlations between those two factors. The results showed that the linear regression models are suitable for predicting the COVID-19 situation (Roseline et al. 2020). Moreover, Mu et al. (2014) analysed a dataset containing Boston suburb house values and used several machine learning methods to make forecasts. The authors used the following algorithms: support vector machine (SVM), least squares support vector machine (LSSVM), and partial least squares (PLS) methods. The results showed that the SVM and LSSVM methods were superior to the PLS method for the nonlinear dataset.

3 Methodology

3.1 Data Processing

The dataset is collected as a combination of two datasets healthcare investment and global COVID-19 statistics. The former contains the attributes of healthcare investment and the level of healthcare in different countries. Its columns include: health expenditure (level of current health expenditure expressed as a percentage of GDP (CE)), share of current health expenditures funded from domestic public sources for health (DE), share of out-of-pocket payments of total current health expenditures (OE), current expenditures on health per capita in current US dollars (CEP), Current expenditures on health per capita expressed in international dollars at purchasing power parity (CPPP), Share of current health expenditures funded from external sources (EE), healthcare workers [Physicians include generalist and specialist medical practitioners (PP), nurses and midwives (NM)], and completeness of specialist workforce (SF). The latter is the COVID-19 global dataset. It contains the daily new infected and death cases as well as the cumulative data for different countries. By using such data, we collected the cumulative data [cumulative cases (CC) and death (CD)] and merged it with the first dataset. Furthermore, the dataset contains the basic information on healthcare expenditure and healthcare workers and the cumulative cases and death for each country. The abbreviations are summarised in Table 1.

Table 1 Abbreviation table

Abbreviation	Description
CE	Current health expenditure (% of GDP)
DE	Domestic general government health expenditure (% of current health expenditure)
OE	Out-of-pocket expenditure (% of current health expenditure)
CEP	Current health expenditure per capita (current US$)
CPPP	Current health expenditure per capita, PPP (current international $)
EE	External health expenditure (% of current health expenditure)
PP	Physicians (per 1000 people)
NM	Nurses and midwives (per 1000 people)
SF	Specialist surgical workforce (per 100,000 population)
CC	Cumulative cases of each country
CD	Cumulative death of each country
CFR	Case fatality rate of each country

3.1.1 Data Cleaning

In the original dataset, some of the data of several countries or regions were missing. For example, there were no corresponding data related to the Faroe Islands and New Caledonia, as two regions in two different locations. Thus, such country/region information was removed in the final dataset. Also, some countries did not have the data for complete infected cases and death registration numbers. As the level of pandemic control largely depends on death-related data, using such data would increase the possibility of building a less accurate model. To deal with the Nan data, three methods are considered. The first one is to abandon all the rows with the missing data, which is definitely not suitable for the current situation. The second method is to use the average of the column to substitute the Nan value. However, it is noted that the development of healthcare is completely uneven across the world. For example, the USA may invest much more in healthcare annually than a poor nation. Thus, replacing the Nan value with the mean figure/value does not correspond with the real situation. The third way is to replace the Nan value with 0. In this study, each feature is normalised, aiming to reduce the error caused by unequal feature range. Moreover, the influence of data cleaning can also be reduced to a minimum.

3.1.2 Data Normalisation

There are two common methods for data normalisation. The first one is Min-Max Normalisation, Min-max normalisation performs a linear transformation on the original data (Patro and Sahu 2015). The general formula for a min-max of [0, 1] is given as: $X' = \frac{X-\text{Min}(X)}{\text{Max}(X)-\text{Min}(X)}$. Also, there is a similar method named Mean normalisation: $X' = \frac{X-Mean(X)}{\text{Max}(X)-\text{Min}(X)}$.

The second method for normalisation is 'Z-score Normalisation'. This method makes the values of each feature in the data have zero-mean and unit variance (Zhang et al. 2021). The general method of calculation is to determine the distribution mean and standard deviation for each feature: $X' = \frac{X-Mean(X)}{STD(X)}$. In this study, we discuss the performance of each normalisation method.

3.2 Linear Regression

Linear regression plays a fundamental role in model analysis, which provides interpretable results on model parameters to find the relation between features (Seber and Lee 2012). A linear regression model is under the assumption of parameter independence (Xiaogang et al. 2012). In order to measure the degree of parameter independence for the variables, we calculated both the Variance Inflation Factor (VIF) and correlation between variables. VIF is calculated by $VIF = \frac{1}{1-R^2}$, where $R^2 = 1 - \frac{SS_{\text{res}}}{SS_{\text{tot}}}$, given $SS_{\text{tot}} = \sum_i (y_i - \bar{y})^2$ and $SS_{\text{res}} = \sum_i (y_i - f_i)^2 = \sum_i e_i^2$. The correlation between variables is calculated using $r_{xy} = \frac{\sum_{i=1}^{n}(x_i - \bar{x})(y_i - \bar{y})}{\sqrt{\sum_{i=1}^{n}(x_i - \bar{x})^2}\sqrt{\sum_{i=1}^{n}(y_i - \bar{y})^2}}$. To benchmark the model, we applied R^2 metrics to score the result model.

3.3 SVR

SVR presents support vector regression, which is an important application of SVM (support vector machine). In this project, we discussed three different SVMs: polynomial kernel SVM, RBF kernel SVM, and linear SVM. The SVM training algorithm creates a model that allocates new samples to one of two categories, making it a binary linear classifier that is non-probabilistic (Cortes and Vapnik 1995). In both the typical Inductive and Transductive situations, SVMs are helpful for text categorisation as they can considerably minimise the demand for labeled training instances (Joachims 1998).

3.4 KNN

KNN is performed by searching the nearest distance or checking the most likely distance between unknown samples and training data sets. In KNN, K represents the number of nearest neighbours to be considered in decision-making. Besides, because KNN is a distance-based algorithm, the classification performance is affected by distance calculation (Yean et al. 2018). Although KNN is widely used because of its low complexity and fast decision-making speed, improving the accuracy of KNN classification is still essential. In order to determine the best K value, it can be selected by checking the data determined by cross-validation (Wettschereck and Dietterich 1994). If k = 1, the error rate of training data is always zero. Thus, the closest point to any training data point is itself. However, when k = 1, the boundary is overfitted (Jerritta et al. 2014). To obtain the favourable value of K, the training and verification set must be separated from the initial dataset. In this regard, it can provide good performance in the range of 1–10 in most cases (Taunk et al. 2019). Well-known distance calculation methods include Manhattan distance ($ManD(X, Y) = \sum_{i=1}^{n} |X_i - Y_i|$), Euclidean distance ($ED(X, Y) = \sqrt{\sum_{i=1}^{n} (|X_i - Y_i|)^2}$), and Minkowski distance ($MinD(X, Y) = (\sum_{i=1}^{n} |X_i - Y_i|^P)^{1/p}$) (Yuvaraj et al. 2014).

The most commonly used distance measure is Euclidean distance (Zhang 2012). Because our dataset is not very huge, KNN has the advantages of short calculation time, high prediction ability, and effectiveness for classification and regression. Therefore, the KNN model is used as a suitable method to predict the minimum healthcare investment required to control COVID-19 effectively. In regression, KNN predicts continuous values, which is the average of its K-nearest neighbour values. Among them, we used cross-validation to obtain the optimal K value and use Euclidean distance to calculate the distance.

3.5 Decision Tree

Decision tree is a tree-structured classifier with three types of nodes. The root node is the initial node that represents the entire sample and can be further split into more nodes. The internal nodes represent features of the dataset, and the branches represent decision rules. Finally, the leaf nodes represent results (Navada et al. 2011). For a particular data point, it completely traverses the entire tree by determining true/false until it reaches a leaf node. The final prediction is the mean of the dependent variable values in that particular leaf node. By iterating, the tree can predict the accuracy of the data point.

4 Results and Discussions

4.1 Linear Regression Analysis

We assumed the model gives no multicollinearity when the value of VIF for all the variables is less than 10 to diagnose VIF's multicollinearity. By this, CEP and

Table 2 Coefficient table for variables

Name	CE	OE	EE	PP	NM	SF
CE	1	− 0.277389	− 0.119709	0.380861	0.440369	0.340688
OE		1	− 0.0119617	− 0.253341	− 0.39544	− 0.196427
EE			1	− 0.578749	− 0.469614	− 0.448343
NM				1	0.685392	0.811243
PP					1	0.641409
SF						1

CPPP values reach 25.765246 and 32.368444 respectively, while others are less than 10. According to Sharma et al. (2021), this issue can be resolved by increasing the dataset size, transforming the multicollinearity features, or removing corresponding variables. In this study, square and square root are performed to transform the variables, however, the value of VIF is still larger than 25 and even reaches 34.643178 for CEP and 44.364177 for CPPP when square root is performed. Table 2 shows that NM, PP, and SF features are highly correlated. Hence, we considered removing the corresponding features while maintaining a relatively high model performance measured by R^2 metrics to maintain the variable independence. The result after removing NM generated the highest score (0.3967).

Table 3 presents the coefficient and the model performance measured by R^2 method. As the dataset is split randomly each time, the value of the coefficient for each variable and the R^2 performance also varies after each execution. Table 4 gives the result of a mean value after executing the model training program multiple times. In the coefficient table, the highest three positive coefficient variables are CE, PP, and SF, with the value of 4.669, 3.117, and 1.342, respectively, where CE is the dominant impact factor among all these features. The results accord with the assumption of Sharma et al. (2021), where they considered the health expenditure of CE and the number of physicians to operationalise the model. Additionally, the coefficient of EE shows a negative correlation with Fatality (the value is − 2.939191). EE denotes the share of current health expenditures funded by external sources. External sources compose of direct foreign transfers and foreign transfers distributed by the government. With a larger value of external health expenditure, it denotes a larger requirement for external funding, which reflects the underdevelopment of healthcare infrastructure (Kiross et al. 2020).

The next step is to validate the result and avoid the influence of the different sizes on the range of each variable. In doing so, the features are normalised using min-max and mean methods. However, it shows a consistent result with the non-normalised table, where the value of SE is − 0.215201, and the value of SF is 0.318. Moreover, we compared the results by using two types of labels, the number of fatalities and the fatality rate. As shown in Table 5, the results present the same correlation between each attribute with the label. The results indicate that they only vary in the degree of proportion.

Table 3 Coefficient of attributes and R^2 score

Feature	*Coefficient*
CE	4.669
OE	− 0.207
EE	− 2.117
PP	3.117
SF	1.342
Metrics	*Value*
R^2	0.474

Table 4 Coefficient table of normalized and non-normalized attributes

Name	Min Max normalization	Mean normalization	Non-normalization
CE	0.137	0.171	4.669
OE	− 0.009	− 0.430	− 0.207
EE	− 0.215	− 0.265	− 2.345
PP	0.022	0.124	3.117
SF	0.318	0.252	1.342

Table 5 Coefficient attributes table of different labels

Name	CD	CFR
CE	4.669	6.916
OE	− 0.207	− 0.702
EE	− 2.117	− 0.225
PP	3.117	3.017
SF	1.342	0.213

4.2 SVR

The first model is linear SVM that introduced as below:

```
linear_svr = SVR(kernel = 'linear')
```

The mean cross-validation score of the default linear SVM is 0.49. Parameter ‘C’ trades off the misclassification of training examples against the simplicity of the decision surface. A low C makes the decision surface smooth, while a high C aims at classifying all training examples correctly. Figure 1 depicts the results of changing the C in linear from 1 to 1000. According to it, the best C value is 10, while the mean score is 0.519. As the figure shows, a better ability to fit the non-linearity is achieved with larger C. Thus, this model gives better choices than the linear kernel.

The second kernel is set as polynomial:

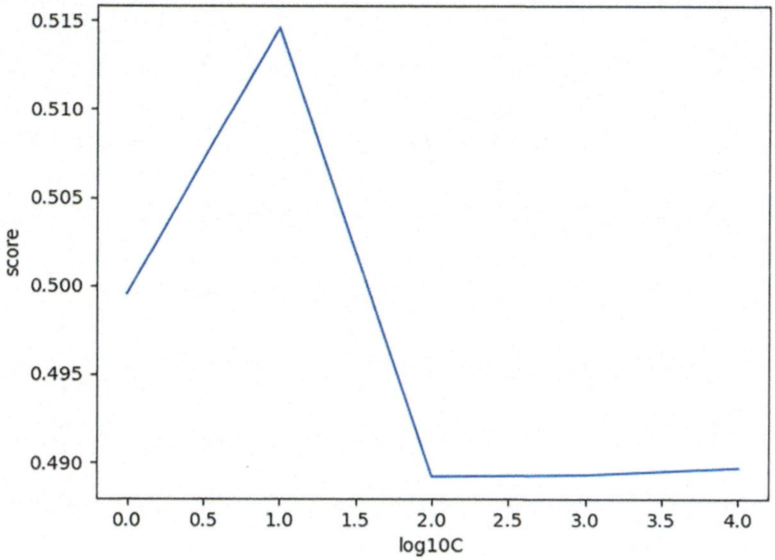

Fig. 1 Linear SVM

```
poly_svr = SVR (kernel= 'poly', C=5,degree=1)
```

According to the results, the mean cross validation score of the default poly-kernel is 0.37, which means the model is not fit well. The SVM with a poly-kernel has two training parameters: C (to control overfitting of the model) and polynomial degree.

As Fig. 2 shows, an optimal result of 0.51 is achieved by changing 'C' (from 1 to 5), and degree (from 1 to 3). The result of the polynomial kernel is similar to the linear kernel, which means they might not be suitable in the current situation.

RBF (Gaussian kernel) is the final kernel setting as below:

```
rbf_svr = SVR (kernel = 'rbf',C=5,gamma=0.9)
```

According to the results, the mean cross validation score of the default Gaussian kernel is 0.54. The SVM with a Gaussian kernel has two training parameters: C to manage model overfitting and Gamma (γ) to control the degree of nonlinearity of the model. As Fig. 3 shows, most of the scores are higher than 0.5 when 'C' and 'gamma' parameters are changed from 1 to 5 and 0.1 to 1 respectively. This means the Gaussian kernel is the most suitable kernel in the SVM model. According to the results, the optimal mean cross-validation score is 0.562.

According to the SVM model training results, the average score for linear SVM, poly SVM, and RBF SVM are 0.38, 0.38, and 0.41 after 50 repetitions.

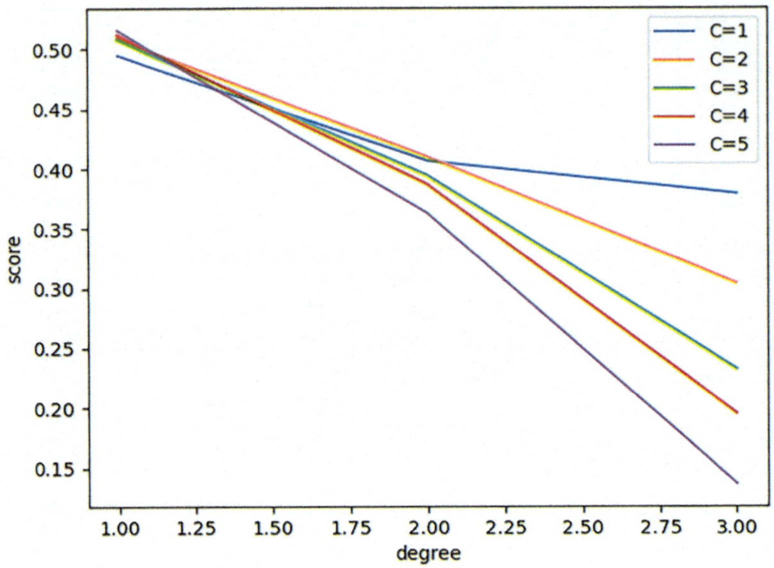

Fig. 2 Ploy SVM

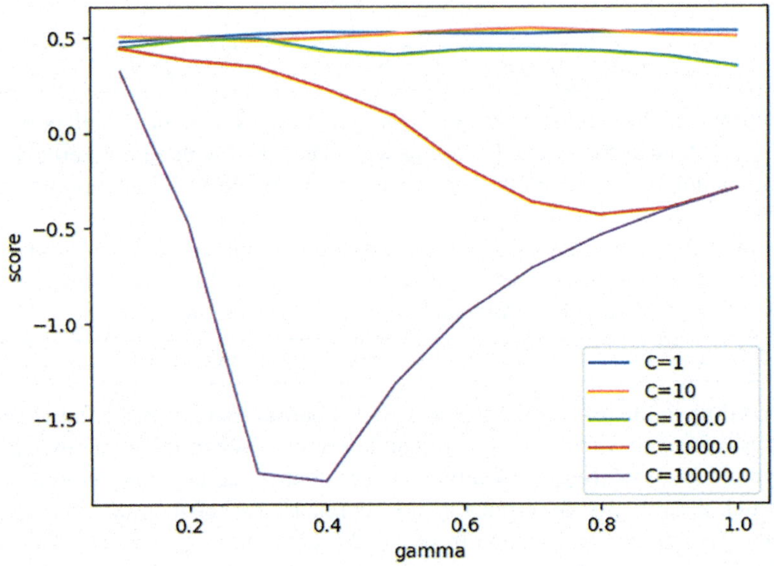

Fig. 3 RBF SVM

Fig. 4 k-score

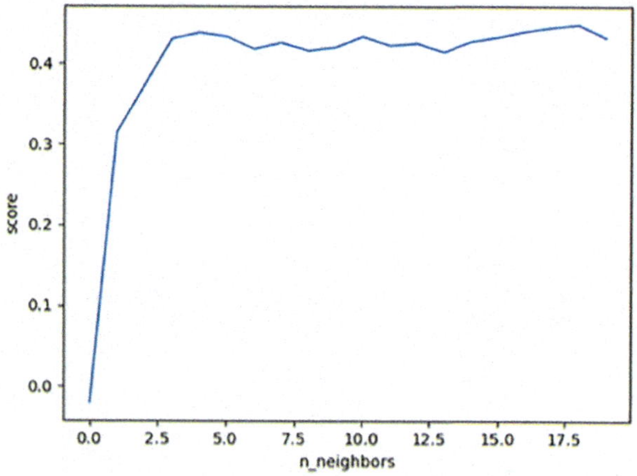

Fig. 5 Times-score

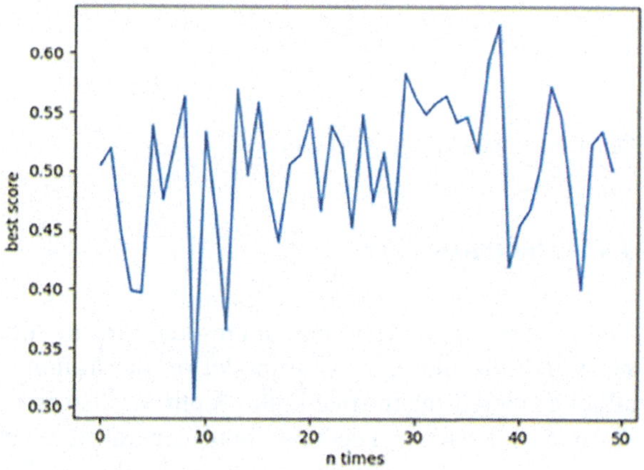

4.3 KNN

The optimal k value needs to b calculated to build a KNN model. For this, a grid-search approach is used to find the best-fitted k value. Figure 4 shows the behaviour of the model when K changes in the range of [1, 21]. In order to improve the accuracy of training, we try to add the weight of the data (i.e., ‘uniform’ to weights = ‘distance’). The grid-search approach is also used to find the optimal p value which is in the range of [0.3–0.5]. Moreover, a normalisation approach is used to map the data to the specified range and make data processing more convenient and faster. At the same time, we aim to simplify the calculation by turning the dimensional data set into pure quantity. By this, the maximum training result reaches 0.655 which gives an optimisation on the original model. Figure 5 shows the overall result trained by 50 repetitions.

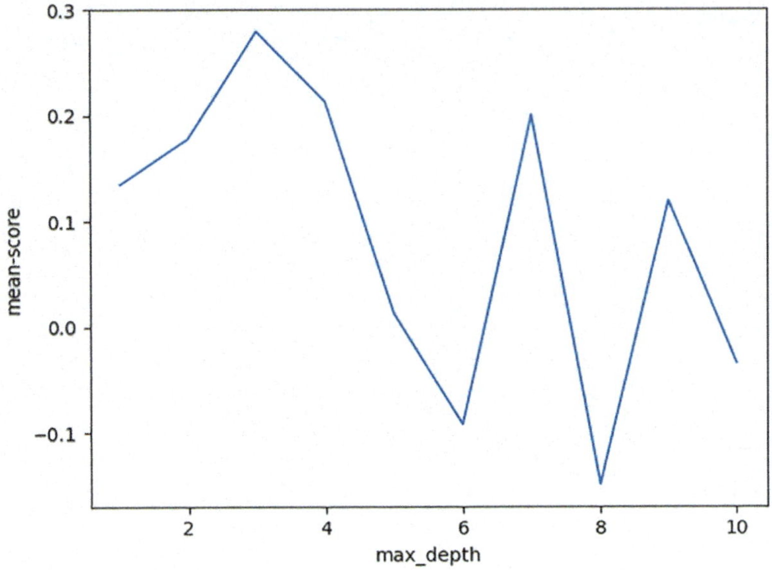

Fig. 6 Maximum depth of decision tree

4.4 Decision Tree

Choosing an appropriate maximum depth for the decision tree model is important when training the regression model on the training set. The literature shows that although increasing the maximum depth can also increase the ability to fit the training set, it may also mean a decrease in its generalisation ability (Jiang et al. 2020). Thus, we use traversal to get relatively good values, which make the max_depth value selection in the range [1, 11]. In Fig. 6, we see that the model mean score takes an extreme value when making a decision when max_depth = 4. Therefore, we perform subsequent training on the basis of max_depth, and achieve the mean score of 0.250. However, the result of the decision tree model is still relatively poor. This is because of the data high variance and instability.

4.5 Discussions

Table 6 shoes the average R^2 score for the machine learning models. According to it, KNN model outperforms SVM and linear regression and gives the best model with an average score of 0.511. This is because KNN is an instance-based model and avoids most of the problems from the linear models, such as independence and multicollinearity. However, the linear regression, with a score close to 0.5, can be still considered a moderate model making the features representative for analysis

Table 6 Results table

Model	Average score after 50 rounds
KNN	0.511
Linear regression	0.474
RBF SVM	0.421
Linear SVM	0.391
Polynomial SVM	0.390
Decision tree	0.250

(Sarstedt et al. 2014). From the results shown in the Linear Regression section, it is clear that the health expenditure (as a percentage of GDP) and the number of healthcare specialists are the most important impact factors contributing to the pandemic control. Moreover, other expenditures, such as external or out-of-pocket expenditures, show an inverse proportion with the pandemic control. Particularly, external expenditures largely impact cumulative death as it reflects the underdevelopment of the level of the healthcare infrastructure of a country (Kiross et al. 2020). In addition, the pandemic control is also affected by other factors, such as the government policy and the learning experience from previous pandemics/outbreaks, which requires a further study on the combination of above factors to generate a more comprehensive model (Sharma et al. 2021).

5 Conclusions

This chapter mainly discusses how healthcare investment affects the control of COVID-19. By comparing and analysing the training results of linear regression, SVR, KNN, and decision tree model, we found KNN is the best-fitted one and gives the best results. For this, an efficient data pre-processing approach is used that uses Min-Max normalisation to process the data and capture the most useful features in the dataset.

However, further investigation and analysis are required as the results are still low. The dataset contains missing attributes. For this, we aimed to delete the data with serious missing attribute values as it cannot reflect the correlation between the data. Moreover, we set the data that is still null after pre-processing to 0 for training. This results in reduced dataset size and lack of samples to train the models. To deal with this issue, we need to classify countries, calculate the average value of relevant attributes of countries at the same level, and replace the null value with the average value. This will improve the model results.

References

Adiga, Aniruddha, Devdatt Dubhashi, Bryan Lewis, Madhav Marathe, Srinivasan Venkatramanan, and Anil Vullikanti. 2020. Mathematical models for covid-19 pandemic: A comparative analysis. *Journal of the Indian Institute of Science* 100 (4): 793–807.

Ardakani, Saeid Pourroostaei, Tianqi Xia, Ali Cheshmehzangi, and Zhiang Zhang. 2022. An urban-level prediction of lockdown measures impact on the prevalence of the COVID-19 pandemic. *Genus* 78 (1). https://doi.org/10.1186/s41118-022-00174-6.

Berwick, Donald M., and Andrew D. Hackbarth. 2012. Eliminating waste in us health care. *JAMA* 307 (14): 1513–1516.

Cheshmehzangi, Ali, Zhaohui Su, Ruoyu Jin, Ayotunde Dawodu, Maycon Sedrez, Saeid Pourroostaei Ardakani, and Tong Zou. 2023. Space and social distancing in managing and preventing covid-19 community spread: An overview. *HELIYON* 9 (3).

Cortes, Corinna, and Vladimir Vapnik. 1995. Support-vector networks. *Machine Learning* 20 (3): 273–297.

Diop, Samba, and Simplice Asongu. 2020. Global health care infrastructure and Africa in times of covid-19: Insights for sustainable development and future pandemics. European Xtramile Centre of African Studies WP/20/073.

Jerritta, S., M. Murugappan, Khairunizam Wan, and Sazali Yaacob. 2014. Emotion recognition from facial EMG signals using higher order statistics and principal component analysis. *Journal of the Chinese Institute of Engineers* 37 (3): 385–394.

Jiang, Shuhui, Haiyi Mao, Zhengming Ding, and Fu Yun. 2020. Deep decision tree transfer boosting. *IEEE Transactions on Neural Networks and Learning Systems* 31 (2): 383–395. https://doi.org/10.1109/TNNLS.2019.2901273.

Joachims, Thorsten. 1998. Text categorization with support vector machines: Learning with many relevant features. In *European conference on machine learning*, 137–142. Berlin: Springer.

John Hopkins University. Mortality analyses. https://coronavirus.jhu.edu/data/mortality. Retrieved March 2023.

Jovanović, Aleksandar, Peter Klimek, Ortwin Renn, Reto Schneider, Knut Øien, Jason Brown, M. DiGennaro, Y. Liu, V. Pfau, M. Jelić, et al. 2020. Assessing resilience of healthcare infrastructure exposed to covid-19: Emerging risks, resilience indicators, interdependencies and international standards. *Environment Systems and Decisions* 40 (2): 252–286.

Kiross, Girmay Tsegay, Catherine Chojenta, Daniel Barker, and Deborah Loxton. 2020. The effects of health expenditure on infant mortality in Sub-Saharan Africa: Evidence from panel data analysis. *Health Economics Review* 10 (1): 1–9.

Kuster, Anthony C., and Hans J. Overgaard. 2020. A novel comprehensive metric to assess covid-19 testing outcomes: Effects of geography, government, and policy response. Medrxiv.

Lal, Arush, Ngozi A. Erondu, David L. Heymann, Githinji Gitahi, and Robert Yates. 2021. Fragmented health systems in covid-19: Rectifying the misalignment between global health security and universal health coverage. *The Lancet* 397 (10268): 61–67.

Leung, Tak Yan, Piyush Sharma, Pattarin Adithipyangkul, and Peter Hosie. 2020. Gender equity and public health outcomes: The covid-19 experience. *Journal of Business Research* 116: 193–198.

Marathe, Madhav, and Anil Kumar S. Vullikanti. 2013. Computational epidemiology. *Communications of the ACM* 56 (7): 88–96.

Mu, Jingyi, Fang Wu, and Aihua Zhang. 2014. Housing value forecasting based on machine learning methods. In *Abstract and applied analysis*, vol. 2014. Hindawi.

Navada, Arundhati, Aamir Nizam Ansari, Siddharth Patil, and Balwant A. Sonkamble. 2011. Overview of use of decision tree algorithms in machine learning. In *2011 IEEE control and system graduate research colloquium*, 37–42. https://doi.org/10.1109/ICSGRC.2011.5991826.

Ogundokun, Roseline O., Adewale F. Lukman, Golam B.M. Kibria, Joseph B. Awotunde, and Benedita B. Aladeitan. 2020. Predictive modelling of covid-19 confirmed cases in Nigeria. *Infectious Disease Modelling* 5: 543–548.

Patro, S., and Kishore Kumar Sahu. 2015. Normalization: A preprocessing stage. arXiv preprint arXiv:1503.06462.

Sarstedt, Marko, Erik Mooi, et al. 2014. A concise guide to market research. In *The process, data, and methods using IBM SPSS statistics*, vol. 12.

Seber, George A.F., and Alan J. Lee. 2012. *Linear regression analysis*. Hoboken: Wiley.

Sharma, Amalesh, Sourav Bikash Borah, and Aditya C Moses. 2021. Responses to covid-19: The role of governance, healthcare infrastructure, and learning from past pandemics. *Journal of Business Research* 122: 597–607.

Su, Xiaogang, Xin Yan, and Chih-Ling Tsai. 2012. Linear regression. *Wiley Interdisciplinary Reviews: Computational Statistics* 4 (3): 275–294.

Taunk, Kashvi, Sanjukta De, Srishti Verma, and Aleena Swetapadma. 2019. A brief review of nearest neighbor algorithm for learning and classification. In *2019 international conference on intelligent computing and control systems (ICCS)*, 1255–1260. https://doi.org/10.1109/ICCS45141.2019.9065747.

Wang, Lijing, Jiangzhuo Chen, and Madhav Marathe. 2020. Tdefsi: Theory-guided deep learning-based epidemic forecasting with synthetic information. *ACM Transactions on Spatial Algorithms and Systems (TSAS)* 6 (3): 1–39.

Wettschereck, D., and T.G. Dietterich. 1994. Locally adaptive nearest neighbor algorithms. *Advances in Neural Information Processing Systems* 1994 (6): 184–191.

World Bank. World development indicators: Health systems. http://wdi.worldbank.org/table/2.12. Retrieved February 2023.

World Health Organization. Who covid-19 global data. https://covid19.who.int/WHO-COVID-19-global-data.csv. Retrieved February 2023.

Yean, Choong Wen, Wan Khairunizam, Mohammad Iqbal Omar, Murugappan Murugappan, Bong Siao Zheng, Shahriman Abu Bakar, Zuradzman Mohamad Razlan, and Zunaidi Ibrahim. 2018. Analysis of the distance metrics of KNN classifier for EEG signal in stroke patients. In *2018 international conference on computational approach in smart systems design and applications (ICASSDA)*, 1–4. https://doi.org/10.1109/ICASSDA.2018.8477601.

Yuvaraj, R., M. Murugappan, Norlinah Mohamed Ibrahim, Kenneth Sundaraj, Mohd Iqbal Omar, Khairiyah Mohamad, and Rajkumar Palaniappan. 2014. Detection of emotions in Parkinson's disease using higher order spectral features from brain's electrical activity. *Biomedical Signal Processing and Control* 14: 108–116.

Zhang, Shichao. 2012. Nearest neighbor selection for iteratively knn imputation. *Journal of Systems and Software* 85 (11): 2541–2552.

Zhang, Zhiang, Ali Cheshmehzangi, and Saeid Pourroostaei Ardakani. 2021. A data-driven clustering analysis for the impact of COVID-19 on the electricity consumption pattern of Zhejiang province, China. *Energies* 14 (23): 8187. https://doi.org/10.3390/en14238187.

Chapter 7
Big Data for Social Media Analysis During the COVID-19 Pandemic: An Emotion Analysis Based on Influences from Social Networks

Abstract Since the beginning of the COVID-19 outbreak, the situation has soon become the most popular topic across social networks worldwide. For almost three years, news and social media were covered with information, updates, and daily/regular reports on the COVID-19 pandemic. When face-to-face activities were restricted, people started to use social networks more than before. The situation led to the empowerment of digital media, such as social media and networks. These platforms became the leading online social hubs for people to express their feelings and record their daily lives during the pandemic. In this study, we analyzed this critical societal change as a major topic, mainly related to the pandemic on Weibo, which is the most popular social media platform in China. The study is focused on one particular context allowing in-depth data analysis related to China's societal impact due to the COVID-19 pandemic. The study also investigates the relationship between COVID-19 trends or topics and public sentiments on social networks. Machine learning is used to verify the correlation between emotion on social media and the COVID-19 pandemic trends. The study concludes with a prediction model using big data for public sentiments on social media.

Keywords COVID-19 · Emotion · Social networks · Weibo · Text mining

1 Introduction

Social networks have become one of the most important ways to quickly share information. Information can be disseminated fast and (often) freely on social networks. The information can be truthful, untruthful, or an opinion of the writer (Riddle 2017). Thus, they can create mixed feelings and emotions when disseminated widely. Nowadays, people use social networks all the time, but such platforms can be abused to spread fake/incorrect information quickly, i.e., leading to misinformation and wrong information dissemination. Information passed by social media can change the belief of an individual or a group, or at least influence their mind or thoughts. Thus, it is

The research work in this chapter was supported by our research team members: Shanglun Wu, Chuyan FU, Xi YU, and Kevin FERDINAND.

S. Pourroostaei Ardakani and A. Cheshmehzangi, *Big Data Analytics for Smart Transport and Healthcare Systems*, Urban Sustainability,
https://doi.org/10.1007/978-981-99-6620-2_7

hard to judge whether the information on social media is truthful or not (Merchant and Lurie 2012); meaning there are two sides of the coin that should be taken into consideration more carefully.

The popularity of social media networks could differ from country to country. Some countries have their own local virtual social networks that have become more popular in recent years. But globally, there is a growing trend of over-reliance on social media networks where people often spend substantial time on the internet or virtual social platforms. In China, there are a few major social media networks or platforms. Weibo is one of the largest social media platforms in China that allows users to share information and get information in their network (Hu 2023). On a monthly basis, it is recorded that Weibo has over 500 million users on its platform. Also, there is a wide range of information exchange between different people and groups on a regular basis. Due to its popularity among people of different generations, it can be said that trending topics on Weibo can influence public opinion to a great extent.

Social media usage and topic trends tend to surge during a major event or incident. This is common when the general public's attention shifts to a particular event. Some examples are natural disasters or a worldwide event, which often attract attention. These events could be positive or negative; but they suggest the importance of the event in people's minds. During such events, people become more active in seeking and receiving the latest information quickly. When the COVID-19 outbreak started and was eventually announced in early 2020, information about the disease spread faster than the virus (Edgar et al. 2020). To some extent, such rapid information dissemination could help the healthcare system to inform everyone about the dangers of the disease. Also, such networks or platforms could be utilised to inform (or even educate) people about methods of dealing with the disease, preventing it, and taking precautionary measures. A research study conducted on Weibo states that more people are discontent about the situation (Fan et al. 2014). Such a finding indicates the critical role of social media for information dissemination in society.

China was identified as the original epi-center of the COVID-19 outbreak, where the disease was first identified. After a few weeks, stringent safety and control measures were imposed in many parts of the country. With lockdown measures imposed in most cities and regions, it was time for people to rely more on social media and their virtual networks. This was during the pandemic's early days when none was allowed to travel outside their homes. There was not much choice but for people to turn to social media as a way to convey their opinions, get information, and use it as a source of entertainment (Liu 2022). As the lockdown was prolonged in some parts of the country, users' social media usage increased substantially. From our data, we can verify that there was a clear increase in social media use during the time the COVID-19 outbreak started. Also, at its peak, it is recorded that people were using social media to post their feelings, opinions, and advice. One of the active social networks was Weibo. During the virus's spread and mutation, many people were infected and under medical control. Under this situation, "epidemic" and "epidemic prevention" have become hot keywords on Weibo, and the heat of the topic was influenced by COVID-19 information (such as increasing cases). Therefore, evaluating the 'how' and 'which' aspect of COVID-19 affects social networks (like Weibo) is essential. Such questions need to be comprehensively considered to avoid some potential social and public opinion problems.

This study's primary goal is to collect data from two main sources of 'Weibo user's posts' and 'China's COVID-19 information'. The data that has been collected is processed by using Spark as a data pre-processing tool to divide each post into words and calculate emotion marks (i.e., senti-mark) for posts according to keywords that are emotionally oriented. This process focuses on finding the potential correlation between posts' emotions and COVID-19 changes through noise reduction and data analysis with Spark. Finally, a model is created to predict future emotion marks according to COVID-19 information, which can help to manage the public's opinion and take measures to assure the public of their safety. This method could be used in the future to avoid the spread of wrong/fake information and help manage (not control) the information that is disseminated to the public. In doing so, we see the potential use of big data analysis in keeping society safe and secure.

The remainder of this chapter is organised as: Sect. 2 focuses on the literature, while Sect. 3 explains the research methodology. Section 4 presents and discusses the results, whereas Sect. 5 outlines the key findings of this research, concludes the discussions, and highlights the future work.

2 Literature Review

Since the beginning of the global COVID-19 pandemic in January 2020, various research has been conducted regarding social media and COVID-19. These studies mainly focused on four topics:

- Hot topics in areas affected by COVID-19;
- Sentimental analysis during the COVID-19 outbreak;
- Public opinion towards COVID-19 lockdown;
- The influence of social media on pandemic control.

In this study, we cover the first three topics as they are relevant to the focus of this research. Results related to these three topics are summarised while we improve the investigation techniques according to the current social media environment and information censorship in China.

2.1 Hot Topic Searches: An Analysis of Information Spread During the COVID-19 Pandemic

Hot topic analysis in COVID-19-affected areas is one of the most popular research topics during the early days of the COVID-19 outbreak. De Santis et al. (2020)

conducted research about detecting and tracking relevant topics among Italian tweets in 2020 (De Santis et al. 2020). This research group used the Twitter API and searched for 35 Italian words related to COVID-19, the hot topics were selected from the texts using the TF-IDF algorithm. They also considered the sequence of words and synonyms in their research and eventually created a word relation map to demonstrate the connection of hot words. This map allowed them to combine the strongly-related single words into different sentences and extract meaningful topics. Another group of researchers from Australia has collected more than 13,000,000 posts using Twitter's API (Yin et al. 2020). Since their dataset was relatively large, they applied the Latent Dirichlet Allocation topic model with Doc2vec in order to classify the posts into different categories and then extract the keywords in each category using TF-IDF. On the other side of the planet, Zhu and his colleagues also used the Latent Dirichlet Allocation topic model in their research about spatiotemporal characteristics of big data on social media (Zhu et al. 2020). Their data was collected from a Chinese data provider Qingbo Big Data Technologies Corporation, which has cooperation with many major Chinese social media platforms such as Tencent, Baidu, and Tiktok. Another group of researchers in China recorded all 4056 Weibo hot searches related to COVID-19 from 31st December 2019 to 20th February 2020 and extracted the hot topics based on word frequency (Zhao et al. 2020). The results from the above four studies have shown that people in Western countries prefer to talk about their safety and protection against the virus, while Chinese netizens often focus on the origin of the virus and fighting against rumours related to COVID-19. This difference in the results could be caused by the different pandemic control policies of the Chinese and Western governments, as well as contextual and cultural differences that influence how people search or access information on the internet.

2.2 Social Media Sentimental Analysis Abroad

Another popular subject is the sentimental analysis of social media posts during the COVID-19 period. In the past two to three years, researchers around the globe have made remarkable progress in this field. Yin and her colleagues used a public library VADER to calculate sentimental scores for their data (Yin et al. 2020). By combining VADER with the LDA topic model, they were able to find the trend of public sentiment during any specific time period. Nemes and Kiss had trained a complex RNN model for sentimental analysis on Twitter posts using a metaheuristic approach derived from K-means and cuckoo search (Nemes and Kiss 2021). They compared their model with the popular sentiment analysis library TextBlob and proved that their method could have better performance on smaller datasets. In September 2020, Khan and his team developed a framework for sentimental analysis of COVID-19 Twitter data based on artificial intelligence (Khan et al. 2020). This framework involves eight components with different functions: data collection, data pre-processing, feature extraction, classification, sentimental analysis, etc. The authors claimed that their framework could highly improve the efficiency of COVID-19 text sentiment analysis; but in reality, it was still just a simple tool based on TextBlob public library. The results of the three previous studies showed that during the COVID-19 outbreak, voices on the internet almost remained neutral, and the number of positive tweets was larger than the number of negative tweets. This is despite the fact that sometimes announcements of new policies lead to relatively large emotional fluctuation among social media users.

2.3 Social Media Sentimental Analysis in China

The network environment in China is a lot different from other countries. In China, people commonly use Weibo as their primary social media instead of Twitter. In this way, they usually cannot obtain information from other countries' social media platforms. In fact, most Western social media platforms are banned or blocked in China (e.g., Twitter, Facebook, Instagram, etc.). Thus, the result of sentimental analysis on Chinese social media turns out to be slightly different from other countries. Zhao and his colleagues divided the COVID-19 outbreak into three stages: outbreak, steady, and decline (Zhao et al. 2020). Then they used the ROST CM6.0 sentiment analysis tool to divide the Weibo posts into three categories of 'negative', 'neutral', and 'positive'. Their result showed that the neutral posts were the majority in all three stages. As the severity of the pandemic reduced, the percentage of positive posts gradually increased. However, even in the decline stage, the amount of negative posts is still 25% higher than positive posts. Wang and his team analyzed the negative sentiments on Weibo using the BERT model, which is a popular NLP model based on transformer (Wang et al. 2020). They also divided the posts into three categories. The result of Wang's team indicated that neutral posts were still the majority, but the number of positive posts was 68.7% higher than negative posts, and this ratio remained the same during the first month of the COVID-19 pandemic.

Another group of researchers from Hunan University focused on the long-term effect of the COVID-19 pandemic on public sentiments (Tan et al. 2021). They collected Weibo posts from 2018, 2019, and 2020, then compared the sentiment of posts before, during, and after the pandemic via Tencent NLP API. Their result showed that positive public sentiments experienced a big decrease in the first two months of the COVID-19 outbreak. It then suddenly increased by 50% in three days when the government announced victory on pandemic control in Wuhan. However, it has continued decreasing since then until all safety and control measures were eventually lifted in early January 2023. One possible explanation for this continuous increase in negative sentiments is that the COVID-19 pandemic has caused severe losses to China's economy.

Aside from traditional sentimental analysis, some research groups focused on specific terms or populations. Lyu and his team studied the use of controversial terms on social media during the COVID-19 outbreak (Lyu et al. 2020). Their research verify that Twitter accounts that often make hate speech and use regional discrimination words could obtain 30% more views than relatively mild accounts. This phenomenon led to an increase in negative emotions on the internet during the pandemic. Liu and his colleagues launched an investigation into the influence of social media on Generation Z during the pandemic (Liu et al. 2021). Using the S-O-R psychological model and data collected from surveys, they found out that the information overload on social media during the COVID-19 outbreak is seriously harming the mental health of younger people.

2.4 *Public Opinions Toward COVID-19 Lockdown*

Apart from the above topics, the change of public opinion towards pandemic lockdown and reopening is also a popular topic among scholars. A group of researchers from the University of Houston investigated the public opinion of US residents toward the lockdown and reopening based on Twitter posts and the TextBlob model (Ahmed et al. 2020). Their result demonstrated a strong relationship between the rise of negative public sentiments and the announcement of lockdown policies, showing that most social media users in the USA were against lockdown measures imposed during the pandemic. Their result also showed that the number of new cases did not have significant effects on public sentiments. For Americans, it was the lockdown that truly mattered the most influencing the public's emotions during the hardships. Similarly, Olagunjul and his team analysed the public opinion toward social distancing and lockdown obligatory in Nigeria (Olagunju et al. Seun Olagunju et al. 2020). The result was the opposite of the United States: 80% of social media users support lockdown policies, and 38.2% think their government should have done more in pandemic control. In China, Han and her colleagues investigated the public opinion on social media toward COVID-19 with the LDA topic model and the Chinese text segment library "jieba" (Han et al. 2020). Their research showed that "Stay at home and take necessary precautions", "Bless and pray", and "Take scientific protective measures" were the three most popular topics in Weibo posts, with a combined percentage of 56.63%. This result indicates that more than half of Chinese social media users appeared to be supportive of lockdown policies in March 2020.

2.5 *Text Emotion Analysis*

Text emotion analysis is to quantitatively analyse the emotion tendency of the text based on algorithms. It is mostly utilized to analyze and classify the expressed emotion from subjective text (Ardakani et al. 2021). With the development of the Internet, numerous users have started to participate in and share valuable comments about hot topics on social media. The content they post online conveys a variety of emotional tendencies, such as happy, sad, agreement, and disagreement.

There are two main methods to analyse the emotion of texts. The first method is to process the text based on a sentiment dictionary and syntactic structure. The text emotion analysis study started in the late 1990s when Riloff and Shepherd constructed a sentiment dictionary based on text repository (Riloff and Shepherd 1997). Later on, more research studies considered the relationship between sentimental words with other special words. Turney used Point-wise Mutual Information (PMI) to expand the positive and negative words in dictionary, and then applied the Integrated Sentiment Analysis (ISA) to the text (Ceron et al. 2016; Peter et al. 2003). Their accuracy in processing general predictive data reached 74%. About the Chinese text analysis, Xu and Lin tried to extract nine semantic features from the texts and construct the Chinese sentimental lexicon (Zhu et al. 2006). Besides, Li and Cao proposed their methods that weight should be given priority in calculating semantic propensity (Dun et al. 2008).

The other method is using machine learning methods like Naïve Bayes or SVM. Pang et al. (2005) (Whitelaw et al. 2005) used machine learning to classify the words to be positive or negative. He attempted Naïve Bayes, Maximum Entropy, and SVM to analyse and compare. As a result, SVM was found to be the best way for it reached 80% accuracy. And in Chinese text studies, Tang et al. (2007) tried to take nouns, verbs, adverbs, and adjectives as the features and use different machine learning methods to analyze and compare. The result showed that with a sufficient training set and appropriate feature selection, SVM and Maximum Entropy showed the best accuracy.

By comparing the two methods, the machine learning methods can extract more accurate semantic features from the text, but it requires abundant training data with manually marked labels. Also, the methods show no obvious advantages in analyzing accuracy compared with those using sentiment dictionary. Thus, we selected the first method to analyze the text emotion in this study. After comparing the emotion dictionaries, the one generated by Chen (2012) was selected. The emotion dictionary he built was specialized for Chinese Weibo content based on a series of famous Chinese emotion dictionaries like Dong's HowNet (http://www.keenage.com) and The Lexicon of Chinese Positive Words made by Shi and Zhu (2006). He first selected benchmark words from a collection of Weibo contents, and then calculated the similarity between candidate words and benchmark words through the SO-PMI algorithm. This approach helps to judge the emotional tendency of candidate words and construct an emotion dictionary belonging to the Weibo field. The core idea of the SO-PMI algorithm is to judge the association degree between new words and benchmark words. If the new word has a higher association with positive benchmark words, it will be marked as positive words, and vice versa. If its association is the same with both positive and negative words, it will be judged as neutral words and will not be added to the emotion dictionary. Its formula is shown below:

$$SO - PMI(word) = \sum_{i=1}^{num(pos)} PMI(word, pos_i) - \sum_{i=1}^{num(neg)} PMI(word, neg_i)$$

Where num(pos) is the number of the positive benchmark words, and num(neg) is the number of the negative benchmark words. As the results,

1. If $SO - PMI > 0$, the word is judged as positive word
2. If $SO - PMI = 0$, the word is judged as neural word
3. If $SO - PMI < 0$, the word is judged as negative word

Since this Weibo emotion dictionary was generated based on 500,000 microblogs, it has great accuracy and applicability for this study. Therefore, it was chosen as the basis for attaching emotional labels to Weibo content in this study.

3 Methodology

3.1 Data Collection

In this study, the original data is collected from the most popular Chinese social media, Weibo, using a Python web crawler. The source code of this web crawler is obtained from an existing open-source web crawler based on Python 3.6.6. However, the original code has several problems because it was written two years ago, so the following modifications have been made in order to improve the effectiveness, efficiency, and memory management: The "format time" function is updated to the newest format from Weibo.

There are limitations to obtaining posts using a web crawler in Weibo. The web crawler is only able to collect 10,000 posts per hour, which is considered slow for big data collection. However, this also prevents Weibo from thinking that the web crawler is attacking the server and blocking the IP. This can be resolved by setting the parameter to sleep for 0.5 ms after collecting 20 crawling operations. This will improve the number of posts gathered while not overloading the Weibo server. The memory management of the original code is poor. It has created a buffer for the output file. However, the buffer is not cleared after each flush operation, so the original program could take up more than 16 GB of memory after running for three hours. In order to solve this problem, the "output to file" function is modified, and the memory overhead is lowered to less than 200 MB.

Moreover, the original code does not have restrictions on the number of data collected per day. This may lead to much unnecessary work because, on some days, there are more than 100,000 posts. A restriction method is added to the code to ensure the web crawler crawls no more than 2000 posts daily. We decided to limit to 2000 posts each day because of time constraints in our data collection. There are many official government accounts on Weibo, and there are also many "marketing accounts" that only re-post the news and exaggerate them for more views and profit. Their posts are not needed in this project because they do not represent people's voices. One common feature of these posts is that they usually have a "[]" around their title because those accounts want the titles to be conspicuous, so we have written a filter to abandon the posts with this punctuation. According to the results, these useless posts take up to 12% of all collected posts.

3.2 Data Pre-processing

To further train the machine learning model on the COVID-19's impact on people's performance on Sina Weibo, the content of each crawled text was labeled first. Since the selected measurement was the text's sentiment degree, all Weibo contents were labeled with their sentiment scores. The whole preprocessing steps are shown in Fig. 1.

Fig. 1 Preprocess steps

Word Segmentation For English text emotion analysis, it is easy to detect each word as there is a blank space between each two words. However, in Chinese, a single character is usually not enough to deliver the meaning, and the smallest meaningful unit is the combination of a few words. Thus, Chinese sentences need an appropriate way to separate sentences into meaningful segmentations. There already exists a lot of libraries to do Chinese word segmentation. In this study, we select one of the best Chinese text segmentation modules, Jieba segmentation (Neutrino 2023), to do the word segmentation. Jieba provides various functions allowing users to customise dictionaries to improve the accuracy. Thus, to do the word segmentation, phrases were separated first by punctuations, and then each phrase was further separated by Jieba word segmentation. Positive words dictionary and negative words dictionary were loaded before doing the segmentation. The separated words are stored in lists.

Delete Stop Words After all words were separated, some of the meaningless stop words needed to be deleted. In this regard, the segmented word list was traversed, and the element detected in the stop word dictionary were removed.

Marking Process Firstly, to mark each sentence's sentiment, each word in the cleansed word list was searched in the sentiment dictionary. If the sentimental word was matched, it's location would be recorded. Since the degree adverb and the privative word always exist in front of the sentimental word in Chinese syntactic structures, after locating each sentimental word, the word between the current sentimental word and the previous sentimental word was searched in degree adverb dictionaries and the privative dictionaries to determine the weight of the semantic feature. The degree adverb dictionary contains different degrees, such as words representing most, more, very, or insufficient, etc. Adverbs of degree found in the corresponding dictionary would add the semantic weight by the corresponding dictionary weight, ranging from 0.5 to 2. If the privative word was found, the semantic of the sentence is reversed by multiplying the weight by -1. What's more, the algorithm also considered the exclamatory sentences. If an exclamation mark is found at the end of the sentence, the weight of the semantic would be extra plus 2 to stress the emotion. In the end, the total score of a sentence would be calculated by adding the result of each short sentence's weight multiplies semantic propensity. Two crawled datasets COVID and COVID-prevention were scored as above, and the score was tagged as 'senti_mark' and appended in the original dataset.

Abnormal Data Process Usually, Chinese Weibo content or word-count is restricted to 140 characters. Users can post different types of content, such as comments, news, or online analysis , which indicates the emotion information in Weibo is complex. In addition, compared with traditional Chinese texts, most microblog expressions are non-written and highly colloquial. Most of them are non-standard and have chaotic sentence structures, which makes it difficult to analyse emotions in Chinese microblogs. From the initial preprocessed results, some abnormal high scores were found. For example, one user repeated numerous of the word "very" in her Weibo content, making its semantic mark's weight extremely large. This was considered as a kind of noise. To avoid the impact of such noise on overall emotional level trends, the absolute value of an emotional score 100 was applied as the threshold to limit scoring range. This means, if the calculated score is larger than 100 or smaller than −100, it will be normalised to 100 or −100.

In addition, the dataset that we have collected must be processed to match each other. Using pyspark sql functions with the data about COVID-19 statistics as the base, this data will be matched with two different datasets. This is the data about COVID-19-related posts in general and data about COVID-19 prevention-related posts. Each of these datasets about COVID-19 must be processed to count the average senti-mark for each day. Then, the senti-mark will be matched to the COVID-19 statistic dataset, thus creating 2 dataset where it contains the COVID-19 statistics with the senti-mark for each day. Dates that are empty or have a value of 0 are deleted as they are considered noise and irrelevant to help predict the senti-mark. This leaves 769 data for COVID-19-related posts and 662 data for COVID-19 prevention-related posts.

3.3 *Data Analysis and Visualisation*

The data analysis part is mostly realised by Spark. As shown in the literature, Spark has high efficiency in data processing operations, so the three datasets were loaded into Spark data frames for the following operations. The first part of data analysis is word frequency analysis. For the COVID-19 data and COVID-19 prevention data, the columns named "content" were selected and kept in lists. After that, the word list was joined into a whole content, which is convenient for word segmentation with jieba library. During the word count, words with lengths less than 1 were dropped, and the stop word collection which named "stop.txt" and used in data pre-processing part was reused here to remove meaningless words. The rest words were sorted by frequency, and the top 30 words were selected for both datasets. Then meaningful words were chosen to form the word cloud to realise data visualisation:

The second part of data analysis and visualisation is to find the correlation between COVID-19 and emotion in social networks. In the data pre-processing part, each content was given a senti-mark to express the emotion. Spark was used to group and sort the posts by day and calculate the average senti-mark for each day. In this way, the datasets got the same x-axis: date. According to the column "publish-time" in COVID-19 and COVID-19 prevention datasets and "date" in the China-COVID-19 dataset, the data frames were joined with Spark. In addition, because there were few posts on the topic of COVID-19 in the early period, the data was selected after "2020-01-17" with a data frame filter function. Then, the matplot library was used to plot figures between dates and different columns belonging to China-COVID-19 to select the attributes that are most likely to affect public emotion. Besides, the selected attributes should not be unique, affecting the accuracy of following machine learning.

3.4 Machine Learning

The machine learning aims to predict senti mark based on local new cases, under medical control, cure rate, and asymptomatic cases. The two datasets that have been processed are fitted on three different machine learning methods. The three models are Decision Tree Regressor, Linear Regression, and Random Forest Regressor. Linear regression (Glasserman 2023) tries to predict output by taking dependent and independent values and making a linear equation to represent how the data will be predicted. Decision tree (Roth 2023) builds regression or classification models in the form of a tree structure. It breaks down a dataset into smaller and smaller subsets while, at the same time, an associated decision tree is incrementally developed. It will make a path of all possible outcomes. Random Forest Regressor (Breiman 2023) is a meta estimator that fits a number of classifying decision trees on various sub-samples of the dataset and uses averaging to improve the predictive accuracy and control over-fitting.

To use the machine learning models, the data that has been pre-processed must be loaded, and added a new feature column based on what classes the model is trying to predict by using a Vector Assembler. Then the data is randomly split into 70% for training and 30% for testing. Then the training data of each COVID-19 dataset is fitted into the three models, which will produce three outputs for each dataset with a total of six model outputs.

Each model gives out RMSE (Root Mean Square Error), MAE (Mean Absolute Error) and MAPE (Mean Absolute Percentage Error). Root mean square error shows how far predictions fall from measured true values using Euclidean distance (C3Ai 2023). Mean Absolute error is the difference between the prediction value and the true value of that observation value (C3Ai 2023). The Mean Absolute Percentage Error is used to measure the accuracy of a model. More specifically, the MAPE is a loss function that defines the error of a given model (DataGy 2023). Thus, RMSE and MAE are able to be counted by using the model's in-built function, while MAPE values need to be counted by using the MAPE formula. The MAPE is calculated by finding the absolute difference between the actual and predicted values, divided by the actual value. These ratios are added for all values, and the mean value is taken.

4 Results and Discussion

The raw dataset contains 480,000 samples, each with 13 features. The features include username, content, device model, like numbers, comment numbers, forward numbers, etc. There is also a "location name" feature, but most samples do not have this feature because this is an optional parameter.

All of the data is collected from a Weibo search, indicating a slight reliability problem. Hundreds of Millions of Weibo are posted daily, and most cannot appear on Weibo searches because the Weibo server cannot index all the posts. Thus, only

Fig. 2 This word cloud is based on COVID-19.csv

the posts that are written by authors with higher Weibo weight, who usually has more than 100 followers and high posting frequency, can appear in search result. However, voices from those people cannot fully represent public opinion.

For data analysis, Fig. 2 represents the dataset collected with keyword "epidemic", which presents keywords like: hope, COVID-19, life, work, and so on. Because of COVID-19, many people's life and work are under effect, and some of them like to share their situation through Weibo and pass hope to the public.

Figure 3 shows the high-frequency word around the topic "epidemic prevention". Isolation policies and physical supply are key operations in COVID-19 control. In

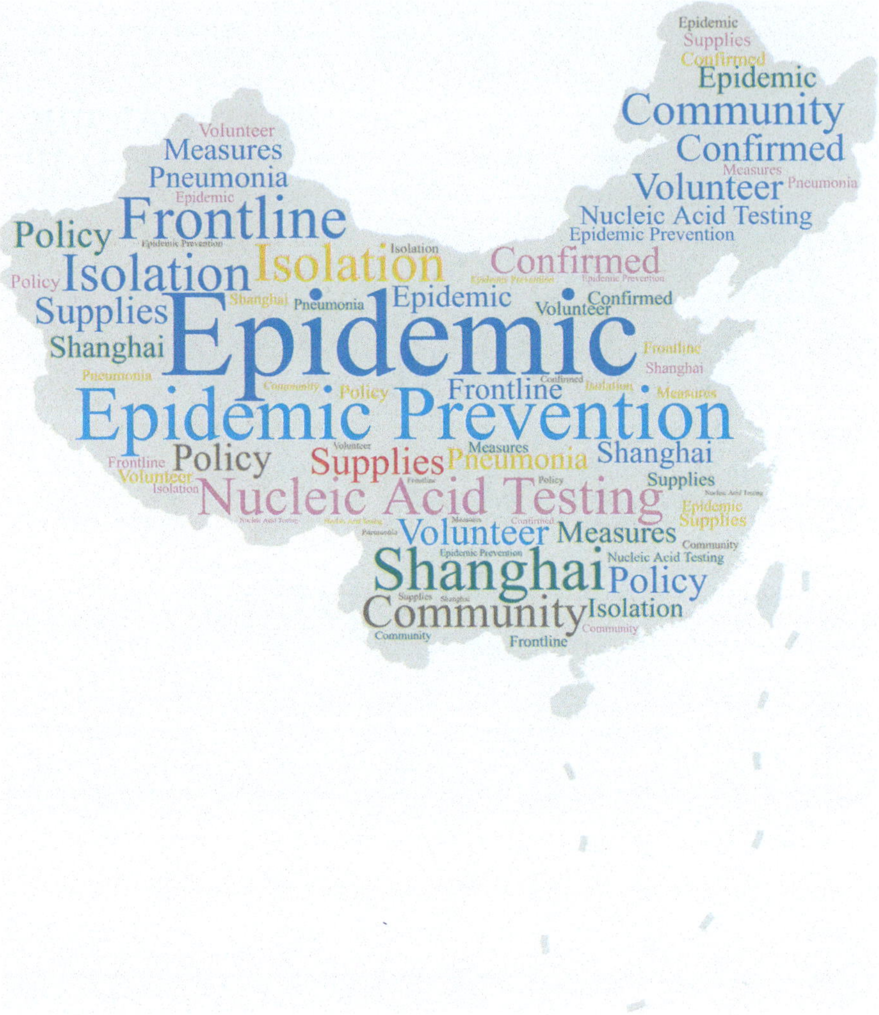

Fig. 3 This word cloud is based on COVID-19 prevention.csv

regards to COVID-19 prevention and control, the government did a relatively good work until March 2022 before the epidemic was widely spread in Shanghai. During this period, many users talked about government's control and prevention policies, and that was the reason for high-frequency of keywords, such as "Shanghai", "Isolation", and "material".

As Fig. 4 shows, "local new cases", "under medical control", "cure rate", and "current asymptomatic cases" were four key features of China-COVID-19 keywords. These are selected as attributes that have a correlation with each other. Considering the small input data frame size, the over-fitting problem is hard to happen.

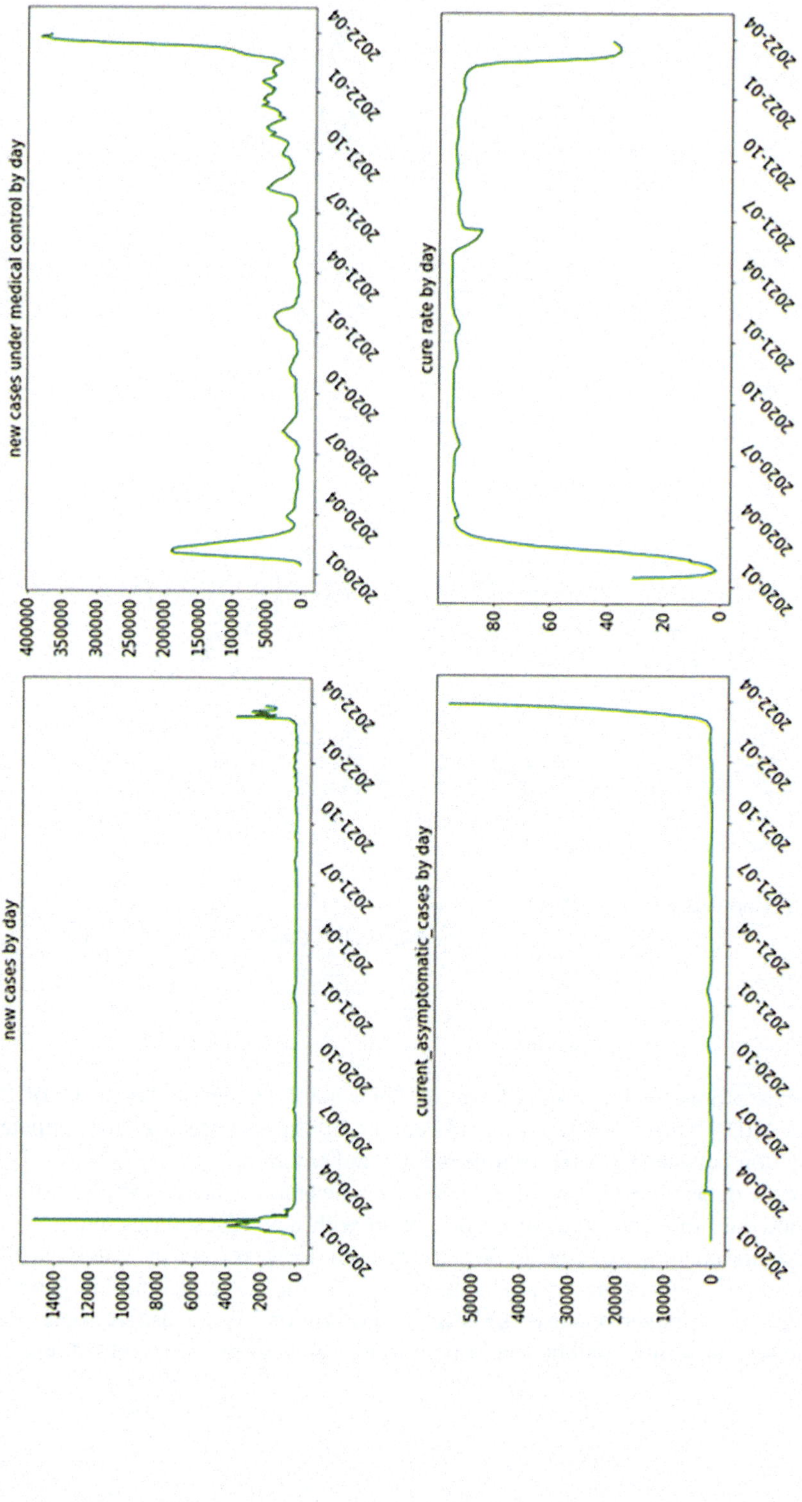

Fig. 4 Figures of attribute selected from China-COVID-19

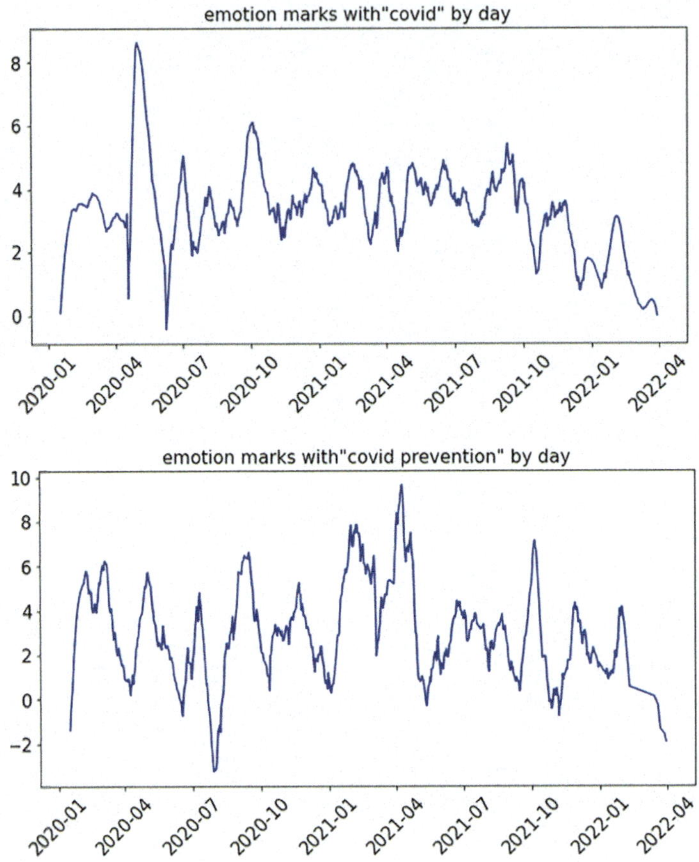

Fig. 5 Senti-mark for two datasets by day

Table 1 Correlation for the COVID-19 dataset

	New cases	Medical control	Cure rate	Asympto-
Emotion	−	−	+	−

Figure 5 presents the graph of average senti-mark change by day for COVID-19 and COVID-19 provided datasets, and Pearson Coefficient was used to calculate the correlation between COVID-19 information and emotion.

For the dataset with keyword "COVID-19", the emotion has a positive correlation with the cure rate and a negative correlation with new cases, cases under control, and asymptomatic cases (see Table 1 for details). However, for the dataset with the keyword "COVID-19 prevention", the emotion has quite a weak correlation with the COVID-19 attributes because "COVID-19 prevention" is a sensitive topic. Based on the results, many content would be considered as aggressive and leading posts and deleted. That is to say, the data sample that is able to be collected is quite small. Also, we note that the Machine Learning results have a significant error mark. Firstly, the senti-mark prediction for the COVID-19 dataset has a significant difference. An average of 3.91 RMSE, with the Linear Regression model having the lowest value of 2.39 and the Decision Tree Regressor having the highest value of 4.71. The MAE average is 1.94, with all three models having almost the same value. The MAPE average value is 184.96, with Random Forest Regressor having the lowest value of 157.35 with Linear Regression model having the highest value of 224.94. The senti-mark prediction for the COVID-19 prevention dataset has an average RMSE value of 5.14, with all three models having around the same value. The MAE average is 3.7, with the three models used having around the same MAE value. The MAPE values of all three models are quite high, with a 319.83 average value.

According to the results, the Linear Regression model has a significant difference compared to Decision Tree and Random Forest Regression for the first dataset. In this regard, we note that the Linear Regression has a higher error mark than the other two machine learning models. However, in the second dataset about COVID-19 Prevention, all three models have around the same error values, which indicates that we are on the right path. The model could be added with more parameters or more data in the future to improve the machine learning result. Taking account of the small dataset and the normalisation of the dataset to be more compact, this margin of error is quite acceptable in representing the average senti-mark for that day.

5 Conclusions

This chapter has brought up a possible correlation between emotion on social media and COVID-19 pandemic trends. We calculated the emotion marks of the posts according to Chinese word processing libraries with the collected data, then presented data analysis and data visualisation in order to find the potential correlation. In the final stage of this research study, we have also developed a prediction model for public sentiments on social media. Compared with other projects, our project focused more on data pre-processing and constructing our own dataset. However, because Weibo was under strict information censorship from the government, a large number of negative posts could not appear in Weibo search results, which affected the reliability and amplified the noise of the data. This might be a possible explanation for the inaccuracy of our machine learning models compared to previous studies in the literature.

In conclusion, we note the result of our model is not as good as initially expected. However, the study contributes to existing and future research through sentiment analysis of social media texts and data visualisation of hot topics. Since our dataset is not large enough (only 244 MB), we did not perform the map&reduce approach with Hadoop and Spark. With more time and larger datasets, future research could further explore various aspects of this proposed scheme, such as collecting more data from different social media platforms, enhancing the quality of the self-constructed dataset, and making more use of Spark libraries. It is important to note that on the 29th of April 2022, Weibo and WeChat (i.e., two primary social media platforms in China), under the pressure of the government, announced that the IP addresses of all users would be visible to the public. This new policy allowed researchers in the future to conduct regional-based pandemic analysis on Chinese social media.

References

Ahmed, Mohammed Emtiaz, Md Rafiqul Islam Rabin, and Farah Naz Chowdhury. 2020. Covid-19: Social media sentiment analysis on reopening. arXiv preprint arXiv:2006.00804.

Ardakani, Saeid Pourroostaei, Can Zhou, Xuting Wu, Yingrui Ma, and Jizhou Che. 2021. A data-driven affective text classification analysis. In *20th IEEE international conference on machine learning and applications (ICMLA), virtual conference*.

Breiman, Leo. 2023. Random forests. https://www.stat.berkeley.edu/~breiman/randomforest2001.pdf. Retrieved February 2023.

C3Ai. 2023. Mean absolute error. https://c3.ai/glossary/data-science/mean-absolute-error/. Retrieved February 2023.

C3Ai. 2023. Root mean square error. https://c3.ai/glossary/data-science/root-mean-square-error-rmse/. Retrieved February 2023.

Ceron, Andrea, Luigi Curini, and Stefano Maria Iacus. 2016. iSA: A fast, scalable and accurate algorithm for sentiment analysis of social media content. *Information Sciences* 367: 105–124.

Chen, Xiaodong. 2012. Jiyu qinggan cidian de zhongwen weibo qinggan qingxiang fenxi yanjiu. *Analysis of sentiment tendency of Chinese microblogs based on sentiment dictionary*. Master's thesis, Huazhong University of Science and Technology.

Cheng, Xueqi, Huifeng Tang, and Songbo Tan. 2007. Jiyu jiandu xuexi de zhongwen qinggan fenlei jishu bijiao yanjiu. A comparative study of Chinese emotion classification techniques based on supervised learning. *Journal of Chinese Information Science* 21 (6): 88–94.

DataGy. How to calculate mape in python. https://datagy.io/mape-python/. Retrieved March 2023.

De Santis, Enrico, Alessio Martino, and Antonello Rizzi. 2020. An infoveillance system for detecting and tracking relevant topics from Italian tweets during the covid-19 event. *IEEE Access* 8: 132527–132538.

Dun, Li, Fuyuan Cao, Yuanda Cao, and Yue-Liang Wan. 2008. Text sentiment classification based on phrase patterns. *Computer Science* 35 (4): 132–134.

Fan, Rui, Jichang Zhao, Yan Chen, and Xu Ke. 2014. Anger is more influential than joy: Sentiment correlation in weibo. *PLoS One* 9 (10): e110184.

Glasserman, Paul. Linear regression. https://www0.gsb.columbia.edu/faculty/pglasserman/B6014/Regression.pdf. Retrieved February 2023.

Han, Xuehua, Juanle Wang, Min Zhang, and Xiaojie Wang. 2020. Using social media to mine and analyze public opinion related to covid-19 in China. *International Journal of Environmental Research and Public Health* 17 (8): 2788.

HowNet. Hownet's home. http://www.keenage.com.

Hu, Shulin. Weibo—How is China's second largest social media platform being used for social research? https://blogs.lse.ac.uk/impactofsocialsciences/2020/03/26/weibo-how-is-chinas-second-largest-social-media-platform-being-used-for-social-research/. Retrieved April 2023.

Khan, Rijwan, Piyush Shrivastava, Aashna Kapoor, Aditi Tiwari, and Abhyudaya Mittal. 2020. Social media analysis with AI: Sentiment analysis techniques for the analysis of twitter covid-19 data. *Journal of Critical Reviews* 7 (9): 2761–2774.

Lerma, Edgar V., A. Verner Venegas-Vera, and Gates B. Colbert. 2020. Positive and negative impact of social media in the covid-19 era. *RCM* 21 (4): 561–564. https://doi.org/10.31083/j.rcm.2020.04.195.

Liu, Hongfei, Wentong Liu, Vignesh Yoganathan, and Victoria-Sophie Osburg. 2021. Covid-19 information overload and generation z's social media discontinuance intention during the pandemic lockdown. *Technological Forecasting and Social Change* 166: 120600.

Liu, Huan. 2022. Official social media and its impact on public behavior during the first wave of covid-19 in China. *BMC Public Health* 428. https://doi.org/10.1186/s12889-022-12803-y.

Lyu, Hanjia, Long Chen, Yu Wang, and Jiebo Luo. 2020. Sense and sensibility: Characterizing social media users regarding the use of controversial terms for covid-19. *IEEE Transactions on Big Data* 7 (6): 952–960.

Merchant, Raina M., and Nicole Lurie. 2020. Social media and emergency preparedness in response to novel coronavirus. *JAMA* 323 (20): 2011–2012.

Nemes, László, and Attila Kiss. 2021. Social media sentiment analysis based on covid-19. *Journal of Information and Telecommunication* 5 (1): 1–15.

Neutrino. Jieba. https://github.com/fxsjy/jieba. Retrieved April 2023.

Olagunju, Olalekan Seun, Obasanjo Afolabi Bolarinwa, and Tesleem Babalola. 2020. Social distancing, lockdown obligatory, and response satisfaction during covid-19 pandemic: Perception of Nigerian social media users. *Advanced Journal of Social Sciences* 7.

Pang, Bo, and Lillian Lee. 2005. Seeing stars: Exploiting class relationships for sentiment categorization with respect to rating scales. arXiv preprint cs/0506075.

Riddle, Jason. 2017. All too easy: Spreading information through social media. *The Arkansas Journal of Social Change and Public Service* 1: 1–5.

Riloff, Ellen, and Jessica Shepherd. 1997. A corpus-based approach for building semantic lexicons. arXiv preprint cmp-lg/9706013.

Roth, Dan. Decision trees. https://www.cis.upenn.edu/danroth/Teaching/CS446-17/LectureNotesNew/dtree/main.pdf. Retrieved February 2023.

Tan, Hao, Sheng-Lan Peng, Chun-Peng Zhu, Zuo You, Ming-Cheng Miao, Shu-Guang Kuai, et al. 2021. Long-term effects of the covid-19 pandemic on public sentiments in mainland China: Sentiment analysis of social media posts. *Journal of Medical Internet Research* 23 (8): e29150.

Turney, Peter D., and Michael L. Littman. 2003. Measuring praise and criticism: Inference of semantic orientation from association. *ACM Transactions on Information Systems (TOIS)* 21 (4): 315–346.

Wang, Tianyi, Ke Lu, Kam Pui Chow, and Qing Zhu. 2020. Covid-19 sensing: Negative sentiment analysis on social media in china via Bert model. *IEEE Access* 8: 138162–138169.

Whitelaw, Casey, Navendu Garg, and Shlomo Argamon. 2005. Using appraisal groups for sentiment analysis. In *Proceedings of the 14th ACM international conference on information and knowledge management*, 625–631.

Yin, Hui, Shuiqiao Yang, and Jianxin Li. 2020. Detecting topic and sentiment dynamics due to covid-19 pandemic using social media. In *International conference on advanced data mining and applications*, 610–623. Berlin: Springer.

Zhao, Yuxin, Sixiang Cheng, Xiaoyan Yu, Huilan Xu, et al. 2020. Chinese public's attention to the covid-19 epidemic on social media: Observational descriptive study. *Journal of Medical Internet Research* 22 (5): e18825.

Zhu, Bangren, Xinqi Zheng, Haiyan Liu, Jiayang Li, and Peipei Wang. 2020. Analysis of spatiotemporal characteristics of big data on social media sentiment with covid-19 epidemic topics. *Chaos, Solitons & Fractals* 140: 110123.

Zhu, Yan-Lan, Jin Min, and Ya-qian Zhou, Xuan-jing Huang, and Wu Li-De. 2006. Semantic orientation computing based on hownet. *Journal of Chinese Information Processing* 20 (1): 14–20.

Zhun, Yinggui, and Jilin Shi. 2006. *The lexicon of Chinese positive words*. Sichuan Lexicon Press.

Chapter 8
Big Data-Enabled Time Series Analysis for Climate Change Analysis in Brazil: An Artificial Neural Network Machine Learning Model

Abstract Climate data is an essential kind of data for humans in the world. Improving the ability to forecast the climate will contribute to the development of many industries, such as agriculture and shipping. In this project, we use the climate data in Brazil from 2000 to 2020. Attributes of the data mainly are date and time, temperature, precipitation, wind speed, and the province in which these data are measured. This study aims to classify these climate data and analyze the changing climate trends in the same province. An artificial neural network is established as the model in this project to implement this objective. The performance shows that this model can complete this classification task.

Keywords Time series analysis · Classification · Climate · Artificial neural network · Brazil

1 Introduction

Climate affects every aspect of human life. Getting Long-term observation data from fixed-point weather stations, such as rainfall, wind speed, relative humidity, atmospheric pressure, solar radiation, air temperature, and dew point temperature, can play an essential role in discovering abnormal, natural weather disaster warnings and climate trends. It will, at the same time, help decrease the cost of production management, improve economic benefits, and even change people's way of life. Considering the climate observation data set provided, we believe it would be a valuable study to predict the observation point of the weather data based on these climate data. There has been little research in this area for a long time, with few theoretical studies and relevant literature. From the theoretical research perspective, more people are needed to study and explore this area, and we believe that our project is pioneering and fills the research gap. Observing the data set provided by

The research work in this chapter was supported by our research team members: Xinyang Du, Jiake Hao, Jiahe Lyu, and Yunuo Zhao.

S. Pourroostaei Ardakani and A. Cheshmehzangi, *Big Data Analytics for Smart Transport and Healthcare Systems*, Urban Sustainability,
https://doi.org/10.1007/978-981-99-6620-2_8

INMET, namely the hourly climate data from 623 weather monitoring stations in Brazil over a period of 20 years, we can find that this data set has the characteristics of spatio-temporal big data.

After a series of analyses, it was found that the climate dataset contained many useless values, such as − 9999, null value, and many interacting values. Those values include Maximum dew point temperature for the last hour and Maximum relative humid temperature for the last hour, from which we need to select representative values for subsequent data analysis. Also, the unrelated and non-representative values should be removed. PCA is an excellent dimensionality reduction method that allows the data to be streamlined and the most representative groups to be extracted from all the useful information. In this case, we extracted 16 groups of labels, among which the most important feature is its high spatial dependence and time dependence. In this dataset, data from different regions interact with each other, and each past moment influences observations at the next moment.

Therefore, compared with the traditional machine learning models, we believe that a machine learning model with time series has more vital learning ability, more comprehensive coverage, and high portability for the time-related problem, and the effect is more prominent in the case of relying on big data. As a result, we introduced the time series model into the model and conducted experiments, and the experimental results verified this view. Our proposed model has higher accuracy than other machine learning models (e.g., decision trees). After analysis, our model is more applicable and stable for such a large amount of data with a long-time span. Hence, this model can be more applicable and accurate in predicting regions.

This chapter is organised as: Sect. 2 reviews the literature—mainly PCA, cross-validation, and time-series models, Sect. 3 describes the research methodology, Sect. 4 presents the results, and Sect. 5 concludes the discussions and address the future work.

2 Related Work

2.1 PCA

Choosing the right training methods when pre-processing a large amount of data is very important and necessary. Common dimensionality reduction methods are LDA, LLE, Laplacian Eigenmaps, etc. Considering the actual situation of this project, we regard PCA as the most appropriate one. Principal Component Analysis (PCA) is a mathematical dimension reduction method that can transform the variables that may be linearly correlated in the data set into new linear unrelated comprehensive indicators through an orthogonal transformation so that the new variables can represent the characteristics of the data in smaller dimensions. As a technology to simplify data sets and reduce data dimension, PCA can retain the features that make the greatest contribution to each other in data sets. The most important aspects of data are retaining low-order principal components (Barnett and Preisendorfer 1987) and ignoring high-order principal components.

PCA has generally performed well in previous projects related to complex weather analysis (Heckenbergerova et al. 2016; Bueso et al. 2018; Mallika et al. 2019; Guo et al. 2020). For example, Wang et al. (2021) innovated TributaryPCA, a distributed streaming PCA algorithm built based on the basic PCA method, in their space weather simulation work. Their results extracted and visualized meaningful changes in simulated exascale data in a time-evolving near-Earth space weather simulation, proving that this approach is efficient and has the potential for significant data prediction. Licciardi et al. (2015) mentioned that they used an NLPCA model for spatiotemporal pattern recognition, a nonlinear extension of the standard PCA. The introduction of PCA significantly improved their prediction accuracy, and the introduction of spatial information could lead to better short-term prediction measurement results for all types of weather conditions, which is consistent with the data model we want to explore. However, Jaruszewicz et al. (2002) found in their study that although the rationality and correctness of the whole could be seen, backpropagation alone was still superior to backpropagation supported by PCA. They believe that the usability of PCA still needs to be improved in the face of meteorological data with high complexity and correlation. In our work, as the data set is enormous, we used PCA to leave the unimportant main components before and after taking a few principal components to the larger variance on behalf of the original variables to eliminate the related influence between the main component. Thus, the difficulty in selecting indicators is greatly reduced, the amount of calculation is decreased, the noise is removed, and the data obtained is considerable and reliable.

2.2 *Cross-Validation*

For the evaluation of the model, after consideration in many aspects, cross-validation is the most suitable validation method for our model. In general training, we split the data into a single training set and a test set. The model learns from the training set and tests its performance in the test set. Then it adjusts itself with the same set. In this process, it is easy to overfit; that is, the resulting model is too corresponding to a particular data set to fit other data or reliably predict future observations. The essence of this phenomenon is that some noise is extracted as the underlying model structure in the adjustment process. Cross-validation is an effective way to avoid this phenomenon. One round of cross-validation involves dividing the data sample into complementary subsets, performing an analysis on one subset, and validating the analysis on the other. After that, to reduce the variability, multiple rounds of cross-validation are performed in different regions, and the results of each round are combined to estimate the model's prediction performance (Kohavi 1995; Cawley and Talbot 2010; Seni and Elder 2010). Cross-validation is usually a better technique to evaluate the performance of neural network models. In our observation, it is widely used in weather-related models and shows good performance (Chandrapala et al. 2021; Raimundo et al. 2021).

In the study of Thonglek et al. (2021), they used the LSTM model to predict future solar power generation through weather forecast data. Finally, they improved the result from 0.750 to 0.805 using K-fold cross-verification. Their data and models were similar to ours, and the improvement was enough to justify the choice. Cerrai et al. (2019) also mentioned in their study that weather and vegetation conditions were used to predict storm interruption areas, and k-fold cross-validation could use strongly correlated adjacent data to conduct model calibration.

2.3 *Time Series Model*

When selecting the model, we found that the LSTM model is very suitable for dealing with problems that are highly correlated with time series. RNN can connect the previous information with the current task. However, as the gap between the relevant information and the target location increases, RNN will gradually find it difficult to learn the connection information. LSTM, as a special RNN that can learn long-term dependence, is designed to avoid this problem. This model is insensitive to gap length and can deal with the possible lag of position duration between important events in time series. Therefore, it is consistent with the situation that we need to analyze the spatio-temporal data for 20 years (Graves 2012). RNN has the following achievements in similar weather-related studies: Yu et al. (2018) proposed a coordinate-based cyclic neural network (RNN) for error correction of numerical weather prediction (NWP) models. Compared with the NWP data from Korea, the proposed rNn-based correction reduces the humidity prediction error by 4.8% and 4.2%, respectively, compared with the uncorrected and simple linear correction predictions.

Mimboro et al. (2021) incorporated RNN into Artificial Intelligence of Things (AIoT) (a New Computing Technology that Combines Artificial Intelligence (AI) and Internet of Things (AIoT) Technologies) for forecasting weather conditions of oil palm plantations. RNN gives good results, and the total average data detection accuracy is 95%. Further development with other AI algorithms is very good regarding the application of AI in the Internet of Things. LSTM has been proven to be a model capable of processing long-term weather data many times (Guo and Lin 2019; Goyal et al. 2021; Lateef et al. 2021; Prathyusha et al. 2021). Hennayake et al. (2021) introduced a model using LSTM to predict temperatures at selected weather stations in Sri Lanka. Their model finally analyzed the Root Mean Squared Error (RMSE), Mean Absolute Error (MAE), and Mean Absolute Percentage Error (MAPE) with extremely low Error values. It shows that LSTM can maintain high accuracy in weather prediction of complex and changeable tropical countries. Yu et al. (2019) introduced the LSTM model to predict solar radiation based on complex weather conditions to solve the gradient disappearance problem in the RNN training process of multi-variable long-term sequence. It is concluded that LSTM is more accurate and effective than RNN. In order to predict traffic flow, Wang et al. (2020) firstly extracted the most influential weather features from weather data through PCA and then used the LSTM model to deal with the complex situation of time characteristics and weather characteristics of traffic flow. Experimental results show that LSTM is suitable for long-term and complex data prediction.

Hachimi et al. (2021) compared the accuracy of machine learning models in their experiments on recommending crops through soil and weather. The results show that the effects of random forest, decision tree, and KNN are much better than those of Naive Bayes. The most significant advantage of LSTM is that it can capture nonlinear relations, but the disadvantage is that it requires a lot of data and professional knowledge to tune parameters. This fits right in with our big data project. In addition, there are many variations of LSTM. For example, the convolution LSTM proposed by Dabhade et al. (2021). They believe that CNN can extract spatial information, while LSTM is very suitable for processing time series data in the deep learning domain. Therefore, they used the combination of CNN and LSTM in the project to detect cyclones in India, and the final results were relatively convincing. Al Sadeque and Bui (2020) used a cascaded LSTM network to predict the weather. The model's short-term memory network learns temporal correlations and shares representations between different weather variables, with the ability to cascade to any prediction length with minimal further training. Finally, the experimental results show that the performance of this model is better than that of deep CNN and conventional LSTM.

2.4 Multi-layer Perceptron

Neural networks are currently commonly used in the field of machine learning and data processing. For example, they can be used for image recognition, speech recognition, etc., predicting the station location using known data. It is a highly parallel information processing system with strong adaptive learning capability. It also does not depend on the mathematical model of the research object, has good robustness to changes in the system parameters of the controlled object and external disturbances, and can handle complex multi-input, multi-output non-linear systems (Lin et al. 2020). The layers are fully interconnected with each other. The input signal is passed forward to the hidden node, and after the radical processing (generally choose S-type interconnection west number). And then, propagate the information forward to the output node layer and finally output the result. If the output layer can not get the desired value, that is, there is a large error, then go to the reverse propagation, the error signal will be returned along the original path after modifying the neuron parameters of each layer so that the error decreases. The BP network is a supervised (tutor) learning method characterised by pattern classification, pattern matching, pattern recognition, and pattern analysis (Quan and Lu 2000).

Considering the complexity of the data set, it is an excellent choice to adopt the MLP model. There are currently many variants of neural networks, such as backpropagation (BP) neural networks, probabilistic neural networks, convolutional neural networks (CNN—for image recognition), long short-term memory networks

(LSTM—for speech recognition), etc. The simplest and most original neural network is the Multi-Layer Perception (MLP), also called Artificial Neural Network (ANN), which is a feed-forward artificial neural network that generates a set of outputs from a set of inputs (Rahmat et al. 2016). The MLP features multiple layers of input nodes as directed graph connections between the input and output layers. MLP uses backpropagation to train the network (Hassib et al. 2019). It is a deep learning method.

The steps of the MLP classifier are as follows. The first step is a basic description of data quality analysis (missing values, outliers, consistency analysis), checking for null values, and deleting all unrelated and useless data. The second step is data transformation. For example, dummy variable processing of discrete features and a different definition of variables to prepare for subsequent predictions can also be used for later direct calls (Teoh et al. 2017). The third step is the type conversion of logical variables, making transitions of data types for easier use. The last step is importing neural networks and new training and testing data. Then the model is automatically generated, and after running the model, users can compare results (Botalb et al. 2018) to see the success rate percentage.

3 Methodology

3.1 Dimension Reduction

After analysis, we found that there were 26 columns in the dataset, of which 17 were related to different climate measurement criteria, and the remaining nine columns were corresponding spatio-temporal data. If these 17 weather data were to be predicted separately, the results would be too tedious and not have the potential of visualization. Therefore, we plan to use Pearson correlation coefficient to find the correlation between each group of data, hoping to find more representative data as the final result of prediction. Pearson correlation coefficient between two variables is defined as the quotient of covariance and standard deviation between two variables. After calculation, the following Fig. 1 and Table 1 are obtained. In Fig. 1, the darker the colour is, the more related the two features are: The final result shows that the highest mean correlation is 0.38. The correlation of these data is very poor, and there is basically no data that can represent the overall trend. So we abandoned this approach and used PCA for dimensionality reduction.

3.2 Time Series Model

As introduced before, the time series model contributes to predicting the climate data because it can recognize and analyze the underlying nature related to the time, such as the periodicity and trend. A famous model used to handle the time-related problem is

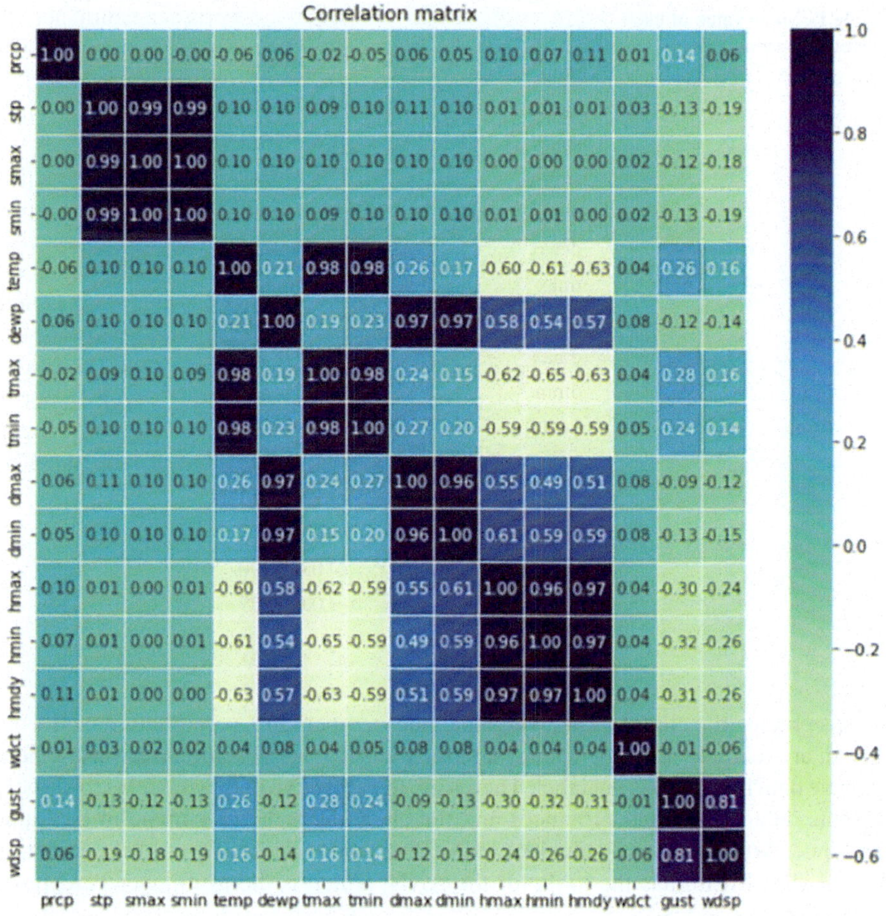

Fig. 1 Correlation matrix result. *Source* The Authors

long short-term memory (LSTM). Both TensorFlow and PyTorch have implemented the LSTM model. However, this project was developed by PySpark and Spark MLlib. The MLlib does not contain the LSTM model or other deep reinforcement learning models, so we used machine learning models in the PySpark and the Spark MLlib, implementing some thoughts of the time series. For data with a degree of periodicity, setting a sliding window can help to find the underlying nature of the data. The value of the dependent variable in the current moment not only depends on the value of the independent variable at this moment but also be related to the value of the independent variables in previous moments. Therefore, the values of previous independent variables should be included in the sliding window for prediction.

An example is illustrated in Fig. 2. The sliding window is from $t - 9$ to t. For this example, to predict the label in the t moment ($X_{t,}$ …, X_{t-9}), the features in the moments from t − 9 to t should be regarded as the inputs to the prediction model.

Table 1 Mean value of each correlation value

Features	Correlation mean
prcp	0.097340
stp	0.271144
smax	0.271528
smin	0.270517
temp	0.166991
dewp	0.374252
tmax	0.160820
tmin	0.173000
dmax	0.378206
dmin	0.371520
hmax	0.180601
hmin	0.166710
hmdy	0.172743
wdct	0.114695
gust	0.067390
wdsp	0.038527

Source The Authors

In this project, the climate data, which includes temperature, precipitation, and wind speed, are features. The province where the meteorological station measuring these climate data is located is the label. Therefore, the objective of this model is to classify climate data in different provinces and the changing trend of climate in the same province. Artificial neural network (ANN) (Fig. 3) is a frequently used classification model. The MLlib has the tools for the ANN, which is the Multilayer Perceptron Classifier.

The weights of nodes between layers are learned from iterative calculation and updating. The updating formula of weights of nodes is formulated below. Since momentum learning is not applied in the ANN in this project, this formula does not include momentum learning.

$$w_{ji} \leftarrow w_{ji} - \eta\delta_j x_i$$

where δ_j is $\frac{\partial E}{\partial(\sum w_{ji}x_i)}$.

In this project, the inputs of the ANN are the dimensionality-reduced climate data, and the outputs of the ANN are the province classification of the input data. To evaluate the trained model, the Multiclass Classification Evaluator in the PySpark is applied as an evaluation model.

The recurrent neural network (RNN) is also widely used to solve time-related problems. The model in this project also can be regarded as an RNN because features

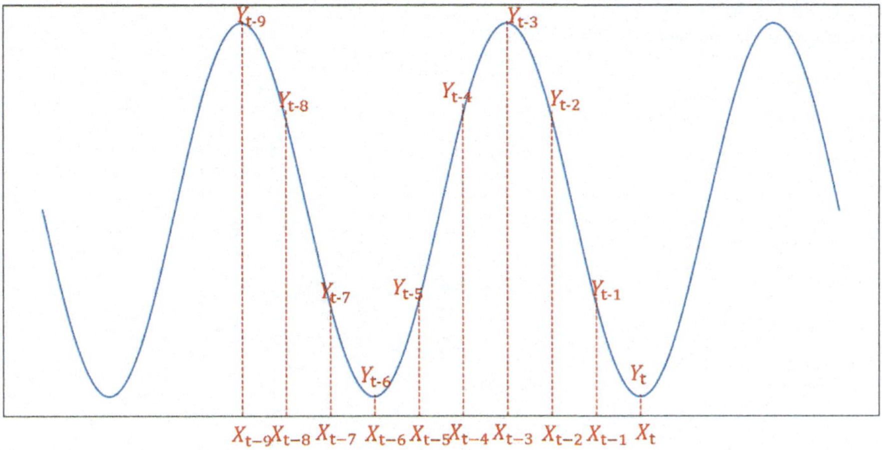

Fig. 2 An example of time-series model. *Source* The Authors

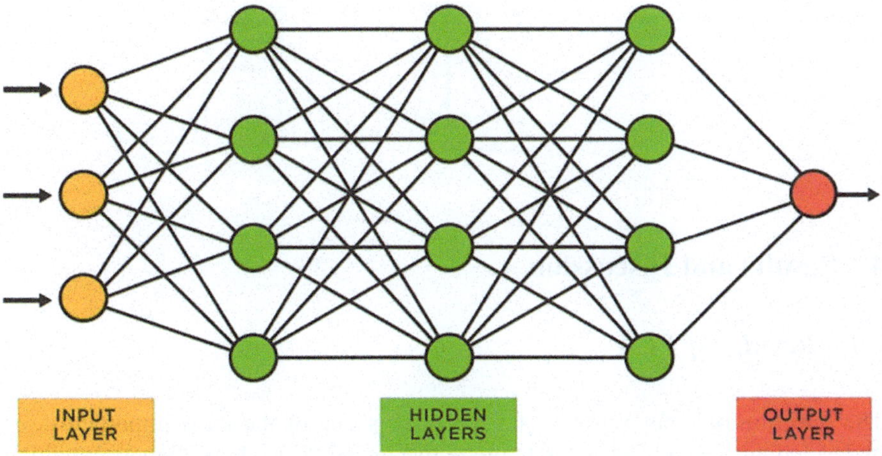

Fig. 3 Artificial neural network. *Source* The Authors

in each moment can be regarded as an input so that features from X_{t-m} to X_t are *m* inputs, as illustrated in Fig. 4.

Fig. 4 Recurrent neural network. *Source* The Authors

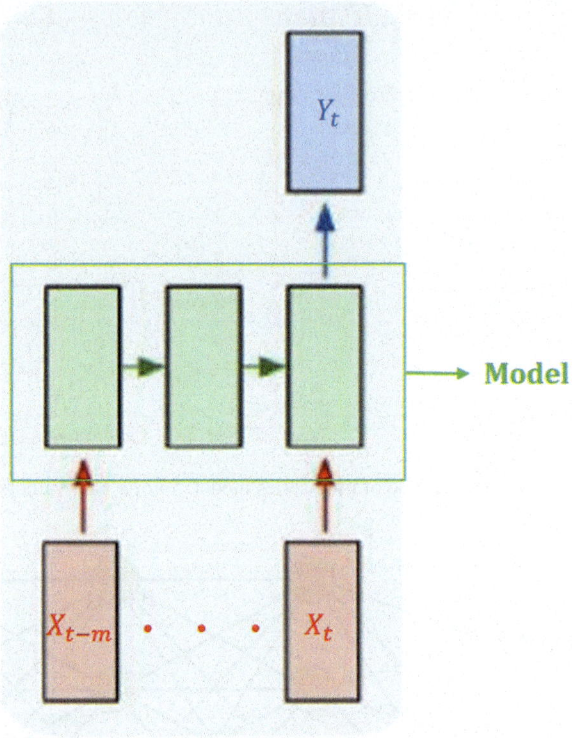

4 Results and Discussion

4.1 Results of PCA

The selection of input variables for a model is one of the most important issues in developing a stable and accurate predictive model. Since local climate conditions receive climatic factors from previous times, we need not only current climatic factors but also climatic factors from past moments to train the model. Only in this way can we improve the accuracy of this model. Therefore, we combined the climate information at the current moment with the climate information from the previous five hours for Brazil. After this, we down-sampled this data using PCA, and the results of the down-sampling were split into the inputs and labels needed for the subsequent MLP model. In applying PCA, we reduced the dimension to 10, and the explained variance ratio is 98.435%.

4.2 Results of Model

In this study, the performance of the model is expressed by the accuracy of the classification. We have used the 'centralwest' data set file for this test and subsequent parameter adjustments, and we use 60% of the pre-processed data as the training set and 40% of the data as the test set. After adjusting the input, we use MLP to classify the data and calculate the accuracy by cross-validation with the following initial parameters:

As shown in Table 2, the initial accuracy was about 0.76, which is not optimal for a binary classification problem, so we used a decision tree to compare the results to see which was better. The decision tree is a simple machine learning model that has the advantage of being fast when testing datasets. It can produce feasible and good accuracy results for large data sources in a short time, especially for a classification model such as our project. The results are shown in Table 3.

As can be seen from the above table, we used 60% of the input data as the training set and used the remaining 40% of the data as the test set for the decision tree. The prediction of the decision tree model was 0.78. It seems that the decision tree model is slightly better than our model, so we decided to adjust the parameters of the MLP model, including the Hyper-parameter, such as the number of neurons in the neural network and the depth of the neural network, the block size of the neural network, and the maximum iterations of the training times of model. This includes adjustments to the non-hyper parameters, such as the number of random seeds, and so on. The results of the various adjustments are shown in the following table. including the number of neurons in the neural network, the depth of the neural network, the block size of the neural network, and the maximum number of iterations of the model training. We also adjusted the non-hyper-parameters, such as the number of random seeds, etc. The results of the various adjustments are shown in the following tables (i.e., Tables 4, 5, 6, 7 and 8).

Although the different selections of those hyper-parameters did not have a large impact on the model, we still had expectations for the after-commission procedure, so we started adjusting the other parameters, such as the depth of the neural network and the number of neurons on each layer. As can be seen from the above three tables, the

Table 2 Primitive parameter of model

Layers	Block size	Max-iterations	Ran seeds	Accuracy
10-32-16-2	128	1000	123	0.761850

Source The Authors

Table 3 The decision tree

Train	Test	Accuracy
60% of input	40% of input	0.777593

Source The Authors

Table 4 The adjustments of 'Block size'

Layers	Block size	Max-iterations	Ran seeds	Accuracy
10-32-16-2	20	1000	123	0.761850
10-32-16-2	40	1000	123	0.755536
10-32-16-2	70	1000	123	0.756400
10-32-16-2	150	1000	123	0.758896

Source The Authors

Table 5 The adjustments of 'Random seeds'

Layers	Block size	Max-iterations	Ran seeds	Accuracy
10-32-16-2	20	1000	20	0.754876
10-32-16-2	20	1000	40	0.748383
10-32-16-2	20	1000	70	0.753324
10-32-16-2	20	1000	150	0.756575

Source The Authors

Table 6 The adjustments of 'Max-iterations'

Layers	Block size	Max-iterations	Ran seeds	Accuracy
10-32-16-2	20	1000	20	0.702422
10-32-16-2	20	500	20	0.746440
10-32-16-2	20	2000	20	0.758887
10-32-16-2	20	3000	20	0.768277

Source The Authors

Table 7 The adjustments of 'Nodes' and 'Layers'

Layers	Block size	Max-iterations	Ran seeds	Accuracy
10-32	128	1000	123	0.026645
10-70	128	1000	123	0.042180
10-32-32	128	1000	123	0.720918
10-70-70	128	1000	123	0.724716
10-110-110	128	1000	123	0.735535
10-32-16-16	128	1000	123	0.749473
10-32-32-32	128	1000	123	0.762119
10-70-70-70	128	1000	123	0.781040
10-32-32-32-2	128	1000	123	0.758368

Source The Authors

Table 8 Other directions results

Data sets	Layers	Accuracy
North	10-70-70-70	0.750271
Northeast	10-32-70-70	0.810633
South	10-70-70-70	0.720918
Southeast	10-70-70-70	0.719731

Source The Authors

adjustment of different parameters does not significantly affect the overall prediction accuracy of the model, except when the maximum number of iterations of the model training is 100, where the accuracy is 70%, which means that the model has not yet completed convergence. We expect the model's accuracy to vary enormously with the following parameters. However, the results are as follows.

The results are as follows. It can be seen that we increased the depth of the neurons from two to five dimensions, respectively, and performed one to three tests with different numbers of neurons in each dimension. However, the adjustment of this hyper-parameter did not significantly affect the model's accuracy in predicting the climate, with the only exception occurring when the depth of the neuron network was in two dimensions, where the model's predictions appeared to be anomalous. After adjusting the model parameters, we started testing the climate data for the other four data sets (north, northeast, etc.). We used the same parameters as before when testing the other datasets because this allowed us to test the stability of the model more accurately. As shown in Table 7, the model was trained with climate data from various regions under the same parameters, and the model's prediction accuracy after the training is generally consistent.

4.3 Results Analysis

The explained variance ratio at around 98% is expressive enough for the original data, so reducing the dimensionality of the data by the PCA technique is reliable. Since the ANN in this project considers the possible intrinsic relationship between the current label with the previous features. However, the decision tree and the ANN performances have a small gap. The ANN, rather than the decision tree, is adopted in this project. For the structure of the model, the performance of the "10-70-70-70" structure is the best, so we use this structure finally in this project. Increasing this maximum iteration before the accuracy converges for the max-iteration parameter will improve the model's performance. However, when the accuracy has converged, which means the model is trained sufficiently; increasing this parameter has no improving effect on the performance.

For the block size and random seed parameters, the performance of the model is not improved. Therefore, these parameters have little effect on the model, especially the performance of the model. This model considers the intrinsic relationship between the province to predict at this moment and climate data at previous moments, but it still has some limitations. Compared to the RNN or LSTM, this method of this model to consider features at previous moments is simple. The sliding window size is subjectively defined, so it may need to recognize the underlying nature of the data. For example, if the sliding window size is smaller than the periodicity, the model may not find this nature. Due to this limitation, we should use sliding windows with an adaptive size. According to the performance of this model on these other four data sets, this model is universal and can be applied widely.

5 Conclusions and Future Work

In this chapter, we train an ANN model with PCA dimensionality reduction and cross-validation. This model also includes the time series by setting a sliding window. For the data in this project, the features are the climate information in Brazil, which includes temperature, precipitation, and wind speed, the label is the province in which these climate data are measured. This study aims to predict this label according to corresponding features. Since climate data is a type of data related to time, not only the features at the current moment but also the features at previous moments can be used to predict the label at the current moment. According to the results of experiments, by adjusting the hyper-parameters and the network structure, the performance of this model surpasses that of some other models, such as the decision tree.

However, this model can still be improved. The sliding window with an adaptive size is expected to be better than that with a subjectively defined size. We can also choose other models, such as the random forest, to implement this classification task. Besides, since PySpark, which is the current applied platform, does not include tools for deep learning (e.g., LSTM, RNN), we will consider using other machine learning platforms, such as TensorFlow and PyTorch. In the future, we will improve the model's performance in these ways.

References

Al Sadeque, Z., and F.M. Bui. 2020. A deep learning approach to predict weather data using cascaded LSTM network. In *2020 IEEE Canadian conference on electrical and computer engineering (CCECE)*, 1–5. https://doi.org/10.1109/CCECE47787.2020.9255716.

Barnett, T.P., and R. Preisendorfer. 1987. Origins and levels of monthly and seasonal forecast skill for United States surface air temperatures determined by canonical correlation analysis. *Monthly Weather Review* 115 (9): 1825–1850.

Botalb, A., M. Moinuddin, U.M. Al-Saggaf, and S.S.A. Ali. 2018. Contrasting convolutional neural network (CNN) with multi-layer perceptron (MLP) for big data analysis. In: *2018 international conference on intelligent and advanced system (ICIAS)*, 1–5. https://doi.org/10.1109/ICIAS.2018.8540626.

Bueso, D., M. Piles, and G. Camps-Valls. 2018. Nonlinear complex PCA for spatio-temporal analysis of global soil moisture. In *IGARSS 2018 IEEE international geoscience and remote sensing symposium*, 5780–5783. https://doi.org/10.1109/IGARSS.2018.8518155.

Cawley, G.C., and N.L. Talbot. 2010. On over-fitting in model selection and subsequent selection bias in performance evaluation. *The Journal of Machine Learning Research* 11: 2079–2107.

Cerrai, D., D.W. Wanik, M.A.E. Bhuiyan, X. Zhang, J. Yang, M.E.B. Frediani, and E.N. Anagnostou. 2019. Predicting storm outages through new representations of weather and vegetation. *IEEE Access* 7: 29639–29654. https://doi.org/10.1109/ACCESS.2019.2902558.

Chandrapala, P., N. Yapage, and M. Mendis. 2021. Daily minimum and maximum temperature forecasting in Sri Lanka: An artificial neural network approach. In *The 2021 5th SLAAI international conference on artificial intelligence (SLAAI-ICAI)*, 1–6. https://doi.org/10.1109/SLAAIICAI54477.2021.9664708.

Dabhade, A., S. Roy, M.S. Moustafa, S.A. Mohamed, R. El Gendy, and S. Barma. 2021. Extreme weather event (cyclone) detection in India using advanced deep learning techniques. In *The 2021 9th international conference on orange technology (ICOT)*, 1–4. https://doi.org/10.1109/ICOT54518.2021.9680663

El Hachimi, C., S. Belaqziz, S. Khabba, and A. Chehbouni. 2021. Towards precision agriculture in Morocco: A machine learning approach for recommending crops and forecasting weather. In *2021 international conference on digital age & technological advances for sustainable development (ICDATA)*, 88–95. https://doi.org/10.1109/ICDATA52997.2021.00026

Goyal, S., N. Mittal, and A. Rana. 2021. Analyzing trend probability and risk estimation of rainfall pattern over Maharashtra. In *The 2021 9th international conference on reliability, infocom technologies and optimization (trends and future directions) (ICRITO)*, 1–5. https://doi.org/10.1109/ICRITO51393.2021.9596217

Graves, A. 2012. Long short-term memory. In *Supervised sequence labelling with recurrent neural networks*, ed. A. Graves, 37–45. Cham: Springer.

Guo, C., and R. Lin. 2019. Analysis of weather and exercise steps based on LSTM neural network. In *The 2019 IEEE 3rd information technology, networking, electronic and automation control conference (ITNEC)*, 474–478. https://doi.org/10.1109/ITNEC.2019.8729294

Guo, S., W. Qiao, B. Chen, and B. Wang. 2020. Prediction and abnormality analysis of climate change based on PCA-ARMA and PCC. In *The 2020 IEEE international conference on networking, sensing and control (ICNSC)*, 1–6. https://doi.org/10.1109/ICNSC48988.2020.9238074

Hassib, E.M., A.I. El-Desouky, E.M. El-Kenawy, and S.M. ElGhamrawy. 2019. An imbalanced big data mining framework for improving optimization algorithms performance. *IEEE Access* 7: 170774–170795. https://doi.org/10.1109/ACCESS.2019.2955983.

Heckenbergerova, J., P. Musilek, and M. Janata. 2016. Sensitivity analysis of PCA method for wind ramp event detection. In *The 2016 IEEE 16th international conference on environment and electrical engineering (EEEIC)*, 1–4. https://doi.org/10.1109/EEEIC.2016.7555687

Hennayake, K.M.S.A., R. Dinalankara, and D.Y. Mudunkotuwa. 2021. Machine learning based weather prediction model for short term weather prediction in Sri Lanka. In *The 2021 10th international conference on information and automation for sustainability (ICIAfS)*, 299–304. https://doi.org/10.1109/ICIAfS52090.2021.9606077

Jaruszewicz, M., and J. Mandziuk. 2002. Application of PCA method to weather prediction task. In *The proceedings of the 9th international conference on neural information processing (ICONIP '02)*, vol. 5, 2359–2363. https://doi.org/10.1109/ICONIP.2002.1201916.

Kohavi, R. (1995). A study of cross-validation and bootstrap for accuracy estimation and model selection. In *The international joint conference on artificial intelligence (IJCAI)*, vol. 14, no. 2. Available at: https://doi.org/10.5555/1643031.1643047

Lateef, K.F., J. Paul, and Z. Jahan. 2021. Weather parameters and heat index prediction applying deep neural networks. In *The 2021 IEEE region 10 symposium (TENSYMP)*, 1–6. https://doi.org/10.1109/TENSYMP52854.2021.9550852.

Licciardi, G.A., R. Dambreville, J. Chanussot, and S. Dubost. 2015. Spatiotemporal pattern recognition and nonlinear PCA for global horizontal irradiance forecasting. *IEEE Geoscience and Remote Sensing Letters* 12 (2): 284–288. https://doi.org/10.1109/LGRS.2014.2335817.

Lin, A., H. Wu, G. Liang, A.C. Tristan, X. Wu, C. Zhao, and D. Li. 2020. A big data-driven dynamic estimation model of relief supplies demand in urban flood disaster. *International Journal of Disaster Risk Reduction* 49: 101682. ISSN 2212-4209. https://doi.org/10.1016/j.ijdrr.2020.101682

Mallika, I.L., D.V. Ratnam, Y. Ostuka, G. Sivavaraprasad, and S. Raman. 2019. Implementation of hybrid ionospheric TEC forecasting algorithm using PCA-NN method. *IEEE Journal of Selected Topics in Applied Earth Observations and Remote Sensing* 12 (1): 371–381. https://doi.org/10.1109/JSTARS.2018.2877445.

Mimboro, P., F. Lumban Gaol, H.L.H. Spits Warnars, and B. Soewito. 2021. Weather monitoring system AIoT based for oil palm plantation using recurrent neural network algorithm. In *2021 IEEE 5th international conference on information technology, information systems and electrical engineering (ICITISEE)*, 283–287. https://doi.org/10.1109/ICITISEE53823.2021.9655818.

Prathyusha, Zakiya, Savya, Tejaswi, N. Alex, and C.C. Sobin. 2021. A method for weather forecasting using machine learning. In *2021 5th conference on information and communication technology (CICT)*, 1–6. https://doi.org/10.1109/CICT53865.2020.9672403.

Quan, Y., and Y. Lu. 2000. Research on weather forecast based on neural networks. In *Proceedings of the 3rd world congress on intelligent control and automation (Cat. No.00EX393)*, vol. 2, 1069–1072. https://doi.org/10.1109/WCICA.2000.863402.

Rahmat, R.F., A. Rizki, A.F. Alharthi, and R. Budiarto. 2016. Big data forecasting using evolving multi-layer perceptron. In *2016 4th Saudi international conference on information technology (big data analysis) (KACSTIT)*, 1–5. https://doi.org/10.1109/KACSTIT.2016.7756069

Raimundo, F., A. Gloria, and P. Sebasti'ao. 2021. Prediction of weather forecast for smart agriculture supported by machine learning. In *The 2021 IEEE World AI IoT Congress (AIIoT)*, 160–164. https://doi.org/10.1109/AIIoT52608.2021.9454184.

Seni, G., and J.F. Elder. 2010. *Ensemble methods in data mining: Improving accuracy through combining predictions*. Part of the synthesis lectures on data mining and knowledge discovery, vol. 2, no. 1, 1–26.

Teoh, T.T., Y. Zhang, Y. Nguwi, Y. Elovici, and W.L. Ng. 2017. Analyst intuition inspired high velocity big data analysis using PCA ranked fuzzy k-means clustering with multi-layer perceptron (MLP) to obviate cyber security risk. In *2017 13th international conference on natural computation, fuzzy systems and knowledge discovery (ICNC-FSKD)*, 1790–1793. https://doi.org/10.1109/FSKD.2017.8393038.

Thonglek, K., K. Ichikawa, K. Takahashi, C. Nakasan, K. Yuasa, T. Babasaki, and H. Iida. 2021. Enhancing the prediction accuracy of solar power generation using a generative adversarial network. In *The 2021 IEEE green energy and smart systems conference (IGESSC)*, 1–6. https://doi.org/10.1109/IGESSC53124.2021.9618702.

Wang, M., T. Shan, C. Chen, and J. Zhong. 2020. Short-time traffic flow forecast with weather characteristics. In *The 2020 international conference on computer communication and network security (CCNS)*, 142–145. https://doi.org/10.1109/CCNS50731.2020.00039.

Wang, Y., N. Klein, S.K. Morley, V.K. Jordanova, M. Henderson, A. Biswas, and E. Lawrence. 2021. TributaryPCA: Distributed, streaming PCA for in situ dimension reduction with application to space weather simulations. In *The 7th international workshop on data analysis and reduction for big scientific data (DRBSD-7)*, 33–39. https://doi.org/10.1109/DRBSD754563.2021.00009.

Yu, C., H. Ahn, and J. Seok. 2018. Coordinate-RNN for error correction on numerical weather prediction. In *The 2018 international conference on electronics, information, and communication (ICEIC)*, 1–3. https://doi.org/10.23919/ELINFOCOM.2018.8330699.

Yu, Y., J. Cao, and J. Zhu. 2019. An LSTM short-term solar irradiance forecasting under complicated weather conditions. *IEEE Access* 7: 145651–145666. https://doi.org/10.1109/ACCESS.2019.2946057.

Chapter 9
Optimized Clustering Model for Healthcare Sentiments on Twitter: A Big Data Analysis Approach

Abstract Social media, such as Twitter, typically stores a large amount of user-generated content regarding different aspects of society. These contents include social events, e-commerce products, healthcare, etc. This chapter proposes a best-fitted clustering method to classify sentiment samples related to healthcare topics. Thus, we examine other clustering models with keyword extraction methods on the real healthcare datasets collected from Twitter. The experiment results indicate that self-organized map model with the TF-IDF extraction method can achieve the best clustering accuracy. Moreover, the optimized model can have great potential to handle large-scale data in real practice.

Keywords Index terms · TF-IDF · Healthcare · Sentiments analysis · Clustering

1 Introduction: Social Media and Public View on Health-Related Matters

Words and sentences represent people's thoughts on services, products, and events via social media. The rapid development of online social networks has brought people's perspectives and opinions into the public's view (Singh et al. 2017). Online users from different social media apps, such as Twitter, Facebook, and Instagram, can share their views and opinions, therefore communicating with readers that grow up in different cultural backgrounds, enriching the overall breadth and depth of topics that are discussed on the corresponding online platform. Moreover, such content would bring an impact on the actual world in the contemporary age. Take the user comments from an online purchasing app, for example; such user-generated comments play a vital role in affecting the purchase decisions of other users. At the same time, those comments serve as an opportunity for organizations and online stores to improve the quality of their products and services. For giant E-commerce corporations like

The research work in this chapter was supported by our research team members: Chenhao Wang, Erfan Wang, Zhe Hua, and Zhengrui Xue.

S. Pourroostaei Ardakani and A. Cheshmehzangi, *Big Data Analytics for Smart Transport and Healthcare Systems*, Urban Sustainability,
https://doi.org/10.1007/978-981-99-6620-2_9

Amazon.com or Taobao.com, by providing the prices, and promotional offers related to the products, they could attract more consumers to make their purchasing decision and leave corresponding reviews demonstrating their sentiments (Wang et al. 2022). After analysing, these corporations can make predictions based on users' buying habits and increase sales performance. Therefore, the sentiment analysis of social media reviews would better help researchers or organizations to predict a user's intention or views regarding the incidents and products and bring actual benefits to society.

More importantly, we believe in the growing importance of social media in maintaining social values, attributes, and factors in modern society. Due to the fact that social media covers fields like social events, healthcare, and e-commerce, a considerable amount of web text enables us to conduct sentiment analysis. Notably, in the area of healthcare, patients would discuss their health conditions on medical sites or forums with respect to their symptoms, treatments, and medications. The patients' on-site visiting experiences to medical hospitals or centers are also posted online (Abualigah et al. 2020). Therefore, it is vital for ill people to learn from other patients with similar symptoms and decide which hospitals to go to or which treatment method is more appropriate for them. In addition, from the medical centers' perspective, it's also beneficial to know the patients' situations and help them solve their problems. Techniques used in the past for analyzing sentiment in healthcare were often carried out in the form of questionnaires and surveys. Still, such approaches took quite a lot of money and time and did not solve patients' problems effectively. Therefore, sentiment analysis tools are needed to exploit the shared documents or comments from the patients. The purpose of sentiment analysis is to cluster and classify the user's comments into different categories to help both the patients and medical centers better.

In addition, this chapter proposes a high-performance clustering model to study Twitter users' sentiments in healthcare. In this study, different clustering models are utilised to group the user-generated healthcare content and evaluate their performance. To obtain characterised features from the text content for clustering, several keywords are extracted with various methods and used to represent the text. The main contributions of the study can be summarised in the followings:

1. A clustering model with a self-organized map (SOM) algorithm has been proposed and optimized to group healthcare sentiments, which can perform high accuracy on genuine sentiment datasets on Twitter.
2. The spark framework has been applied in data loading and normalization, which improves the ability of the clustering model in large-scale processing data.
3. The effectiveness of applying TF-IDF and TextRank in keyword extraction of healthcare sentiment analysis has been explored. The experiment results indicate that the TF-IDF method is more suitable for keyword extraction of short text.

This chapter is structured as follows. Section 2 focuses on the related works, Sect. 3 presents the research methodology—mainly feature extraction and the clustering algorithm, Sect. 4 presents and discusses the experimental results, and Sect. 5 outlines the critical points of this research and highlights the issues that should be addressed as further work.

2 Literature Review on Sentiment Analysis in Healthcare

Related sentiment analysis carried out in the field of health-care should be elaborated because this paper focuses more on medical-related areas. Moreover, healthcare is a data-rich industry (Gupta and Kohli 2016), in which the center of this area focuses on the patients, their choice of the hospital, as well as preferred treatments, is given. To analyse such a huge amount of datasets, techniques in big data are needed. According to Gupta and Kohli (2016), big data comprises three categories of data types: Structured, Unstructured, and Semi-structured.

Structured data is termed a form of organized data that follows a predefined format. Query and reports on predetermined data types were the main goals of using structured data. Unstructured data might demonstrate a complex structure that might not follow a predefined pattern (Pat Research 2020). For semi-structured data, it depicts a schema-less or self-describing structure. It represents structured data with tag elements for the separation of semantic elements. Moreover, hierarchies and fields will be generated by semi-structured data. Therefore, such data types conform to different structures of data models that are in the relation databases (Pat Research 2020). In this regard, big data consists of five different classes of information (Raghupathi and Raghupathi 2014; Chluski and Ziora 2015):

- Social media and Web data: live or interactive data from social media like Twitter, and Facebook that contain healthcare-related data;
- Sensor data: information from meters, and sensors;
- Transactional data: Billings in healthcare centers are important for semi-structured data as well as unstructured data;
- Biometric data: retina, and fingerprint scans belong to this category;
- Auto-generated data: paper documents, or emails that are generated by systems.

Challenges that might be faced by big data in Healthcare (Belle et al. 2015) are also identified in five main categories:

- Not able to perform information exchange across different settings due to failure to take in structured data;
- Knowledge inference gives various patients information;
- Lack of information exchange;
- A precise comprehension of the clinical notes under the correct context;
- Pre-processing of a huge amount of image data and extract out useful information;
- Privacy or security concerns.

The main goal of Big Data Sentiment Analysis is to improve the quality of medical service to better help patients. Such a method also benefits healthcare fraud detection by organizing specific health populations. Moreover, by aggregating patients' experiences, the health center could present a more profound understanding of the healthcare domain, and by improving the service quality, patients' perceptions and satisfaction would increase. A mixed polarity detection approach is probability suitable for sentiment analysis (Bahrainian and Dengel 2013). The following part will introduce more intricate features of big data sentiment analysis processing steps, including keyword extraction and clustering methods used.

2.1 Keywords Extraction Methods

The extraction of correct information based on user requirements relies on keywords. However, numerous books, articles, and comments are published or posted every day, which makes it hard to follow through with the whole text materials. Therefore, reliable text extraction methods are needed, which could provide an accurate content representation of the original text, and effective keywords play a vital role in this process (Kaur and Gupta 2010). In addition, the semantic meaning of a document can be summarized by keywords because they are the fundamental building blocks. Applications can then take advantage of keywords in areas such as auto-indexing, content summarisation, clustering, classification, data visualisation, etc. (Bracewell et al. 2005). Based on Kaur and Gupta (2010), current methods in Automatic Keyword Extraction is split into four classes:

(1) ***Statistic Approaches***

Methods that fall under this category do not need additional training data. Instead, the statistical characteristics are then utilised for keyword identification based on the text content. Auto-indexing, based on NGram and domain-independent statistical information, was proposed by Cohen and Sutton (2003). In addition, methods like TF-IDF (Guo and Yang 2016), PageRank (Brin and Page 1998), and co-occurrences of words were also deemed as statistical approaches (Zhang 2008). The TF-IDF method and PageRank method will be explained a little bit more. Term Frequency-inverse Document Frequency (TF-IDF) is a commonly used weighting technique for information retrieval and data mining by evaluating the importance of a word to a document in a set of documents or a corpus (Rajaraman and Ullman 2011). The importance increases in direct proportion to the number of times a word appears in the document. However, it decreases inversely to the frequency at a word appears in the corpus. Other method is TextRank, which is a graph-based ranking model for text processing and especially NLP. It is derived from Google's PageRank (Brin and Page 1998) which was used in citation analysis, social networks, and typically the analysis of the World Wide Web.

(2) ***Linguistics Approaches***

Linguistic methods mainly exploited features buried in sentences and documents, common methods like lexical analysis (Hanks 2013), syntactic analysis (Kohler 2012), discourse analysis (Wodak 2011), and more.

(3) ***Machine Learning Approaches***

From the aspect of machine learning speaking, keyword extraction can be viewed as a supervised learning process. Such a method would first model the keywords from training documents and then predict the keywords from giving testing documents. Common machine learning models used for keyword classification were Naïve Bayes (Uzun 2005), Support Vector Machine (SVM) (Hu et al. 2018), and Decision Tree (Navada et al. 2011).

(4) ***Other Approaches***

Other approaches related to the extraction of keywords comprised combinations of previous methods mentioned above. But also include some basic heuristics, for example, length, position, or HTML tags. In addition, the above methods covered only handle single documents, but the scale can be increased according to the corresponding fitness of the method. The selection of keywords is normally based on finding the noun phrase inside the document, as those noun phrases consist of vital indications of the original text document. Moreover, linguistic features (Hu and Wu 2006) and informative features (Alguliev and Aliguliyev 2005), like highlighted words, were considered when the choice of keywords was performed. Query-focused keywords were also counted as candidate keywords (Ouyang et al. 2011). For single document extraction, methods like co-occurrence (Ohsawa et al. 1998; Matsuo and Ishizuka 2004) and machine learning (Turney 2000; Hulth 2003) were utilised. For topic detection, keyword clustering was used (Wartena and Brussee 2008). Moreover, history or query frequency-based keywords clustering (Onoda et al. 2008) was also one of the methodologies adopted.

2.2 *Sentiment Analysis Clustering Models*

To better analyse the sentiments buried under documents more reasonably and automatically, unsupervised learning methods came to play. When the keywords extraction process is finished, clustering models should then be applied to those features. The discovery of natural groups was the main goal of the clustering process. Those groups were then used to depict the classes inside the original documents (Salton and Buckley 1988). The most referenced algorithm for clustering was the basic k-means (Hartigan 1975), which did not need prior knowledge about the class of a document, and no training process was required. In other words, the characteristics of k-means could be free from human intervention and, therefore, save time. The feasibility of using clustering in sentiment has been proven by similar research conducted by Hatzivassiloglou and McKeown (1997). However, only the authors can determine adjectives' semantic orientation. In the work carried out by Su and Markert (2008), the proposed method aimed to map implicit expressions that were deemed to be sentiment words to corresponding explicit aspects. In the study by Andrzejewski and Zhu (2009), domain knowledge or constraints were used to guide the modeling of the topic and, therefore, to produce better clusters. Moreover, Zhai et al. (2011) developed an algorithm to cluster product features. Recently, dynamic sentiment analysis was also considered in the sentiment clustering task (Zhu et al. 2014).

In addition, commonly used methods for clustering were:

- K-Means Clustering: K-means is one of the simple unsupervised machine learning techniques for clustering. The input features will be divided into K clusters, and each sample will be assigned to the closest cluster. After that, the center cluster will be recomputed and updated for the next iteration (MacQueen 1967);
- Partitioned Clustering Algorithm: In this algorithm, the solitary partition of the sample is required instead of the cluster structure (Guha et al. 2003). Take an example, the dendrogram indicated by a series of level methods;
- Fuzzy C Means Clustering Algorithm: For this algorithm, the distances between the cluster center and the sample is utilised to delegate the membership function to each data point for each cluster center;
- Self-organized Map (SOM) Clustering: The self-organising map (SOM) is inspired by the neurological structure (Asan and Ercan 2012) and is often deemed to be a helpful tool for the analysis of data. There are several kinds of SOMs, but the most common kind, named Kohonen network (Kohonen 1982, 2001), is a special class of artificial neural networks that are applied in clustering and visualisation tasks;
- Spectral Clustering: Spectral clustering is an algorithm that evolved from graph theory and has been widely used in clustering later. The main idea is to consider all data as space points that can be connected by edges.

3 Methodology

3.1 Feature Extraction Algorithm

Term Frequency–Inverse Document Frequency (TF-IDF): The main idea is that if the frequency of a word or phrase in one article is high, and it rarely exists in other articles, then this word can be considered a good feature to distinguish article categories. Therefore, it will become suitable for the classification (Guo and Yang 2016). TF means term frequency, while IDF means inverse document frequency.

The advantage of TF-IDF is its simplicity. However, the shortcoming of it in keyword extraction is also obvious. It relies heavily on a corpus that needs to be high quality and consistent with the text to be processed.

TextRank: The major adaption of TextRank compared to PageRank (Brin and Page 1998) is that the "strength" of the connection between two vertices (keyword or phrases) is incorporated into the model as a weight to the edge linking the two vertices (Mihalcea and Tarau 2004).

The advantage of TextRank is that it inherits the PageRank (Brin and Page 1998), and relationships between text terms can be utilized more. However, the result is greatly affected by word segmentation and text cleaning; that is, the retention of some stop words might directly affect the final result.

3.2 Clustering Algorithms

K-Means Clustering: K-Means is an unsupervised machine learning algorithm that is based on numerical inputs. Because of its simple yet effective characteristics, K-Means is adopted widely to solve practical problems and can yield good results (Na et al. 2010).

The K-Means clustering also has disadvantages, such as the cluster centers might be affected by the outliers, and the outliers will have their centers. Moreover, as the number of dimensions of the dataset increases, the similarity measures based on numbers will finally converge at a constant value.

Self-organizing Map (SOM): Each training sample works as a different stimulus signal for the SOM neural network that activates the corresponding neuron. Therefore, different samples will fire up different neurons. Moreover, due to each neuron's distinct and special characteristics, various samples can be effectively recognized by the neural network. In other words, the purpose of clustering is reached because samples with similarities activate similar sets of neurons. In this part, the neuron distribution and learning process of the SOM algorithm will be introduced in detail. The SOM network comprises two layers, the input layer, and the output layer, as shown in Fig. 1.

The output layer is called the 'Kohonen layer', which visualises the low dimensional visualization of the data. The input layer is directly connected to the output layer. Moreover, the nodes located in the input layer stand for the features, and those features are represented by an input vector, $x = (x1, x2...xm)$. The size or the number of nodes in the output layer denotes the number of clusters.

Learning Process: The learning processes for the SOM algorithm can be divided into the following part:

- Initialisation: Choose dimension and size of the output space, and normalize the training data. Assign random weights to weight vectors $W_i(0)$. Define neighborhood function h_{ci}, learning rate $\alpha(0)$, size of neighbors $\sigma(0)$, and max iteration value T.
- Sampling: Randomly selection of the training data.
- Similarity Matching: Calculate the Euclidean Distance between input vector and the output vector and find the best matched node $n(t)$ using the Eq. 1 below:

$$n(t)_i = \arg\min \|x(t) - w_i(t)\| i = 1, 2, \ldots, n \tag{1}$$

- Weight Updating: Update the weights for both the winning node as well as its neighbors with the update Eq. 2:

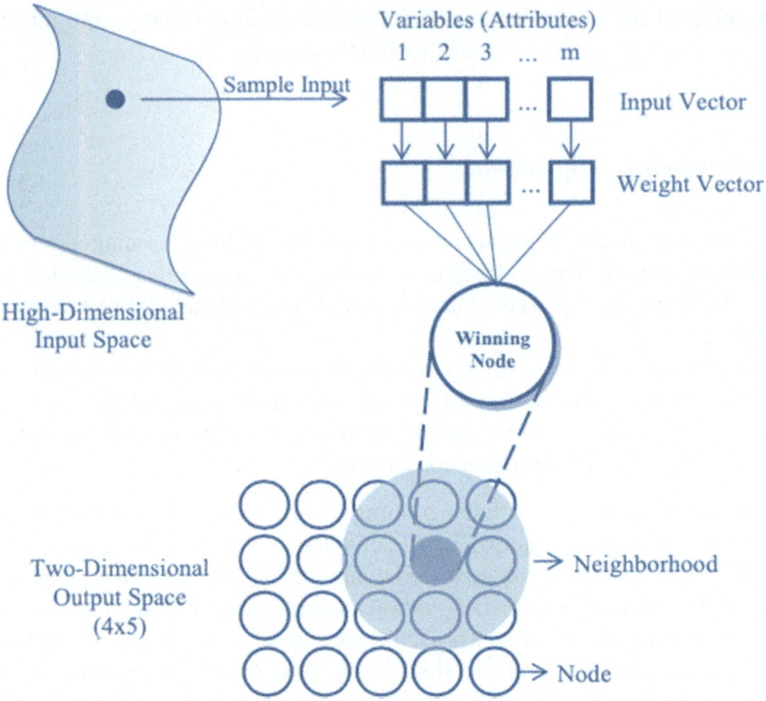

Fig. 1 Graph illustration of self-organizing map. *Source* Adapted by the authors, originally from Asan and Ercan (2012)

$$w_i(t+1) = w_i(t) + \alpha(t)h_{ci}(t)[x(t) - w_i(t)] \tag{2}$$

From the winning node's aspect, the neighboring function $h_{ci}(t)$ is set to 1.

- Parameter Adjustment: Update t = t + 1, adjust σ and α.
- Continuation: Jump back to step 2 until stopping iteration T is reached.

Spectral Clustering: The edge weight between two farther-apart points is lower, while the edge weight between two closer points is higher by cutting the graph composed of all data points to make the sum of edge weights between different sub-graphs as low as possible, while the sum of edge weights within the sub-graphs as high as possible, to achieve the purpose of clustering. The way of representing the data is better using the similarity graph. There are three commonly used methods to compute the similarity graph: ϵ-neighbourhood graph or k-nearest neighbor graph, or fully connected graph.

4 Experiments and Results

4.1 Dataset

The datasets (from: https://archive.ics.uci.edu/ml/datasets/Health+News+in+Twitter) were collected in 2015 using Twitter API. These datasets contain health news from more than 15 major health news agencies such as BBC, CNN, and CBC. The dataset of CBC is used in the experiment since our experiment requires massive manual work of assigning a label. We first apply the keyword extraction methods on the dataset to build a hot-keyword set. To explore the effect of different numbers of clusters on the performance of clustering models, the hot-keyword set will be split into m classes, for which there are m different labels correspondingly. Each instance will be defined with the label n if it contains the most frequent hot keywords belonging to the class n. Therefore, the experiment datasets contain 2000 samples taken from the CBC dataset for the training set and 200 instances for the validation and testing of the clustering models. However, every dataset is assigned a different number of labels from 3 to 8.

4.2 Experiment and Configuration

As shown in Fig. 2, the experiment starts by loading the labelled samples from the dataset using spark frameworks, a fast and general-purpose engine for large-scale data processing. Next, all data samples will be normalized, including converting all letters into lowercase, expanding abbreviations, and removing special characters, punctuation symbols, multiple spaces, single letters, and digits. Some extremely common words, such as 'the', 'an', and 'in' are of little value in selecting keywords from the text. Thus, these stop words are excluded with the help of the stop-word remover in the spark package. Moreover, we transform all words into its stem correspondingly to ensure each extracted keywords are discriminative. After the normalization, the data samples will be split with a ratio of 10:1 for the model training and testing, respectively. In the training stage, the feature extraction will be conducted on those training samples using two keyword extraction methods, namely, TF-IDF and TextRank, to form a vocabulary of hot keywords and construct fix-sized feature vectors with TF-IDF values or one-hot representation of keywords, respectively. Because the data samples are short text, the TF-IDF method selects one keyword if it appears in more than 0.5% of total training samples, while the TextRank method only selects two keywords for each sample with a small window size. And then, three clustering models, that is, K-Means, SOM and Spectral Clustering, are used to fit those feature vectors. In the testing phase, each testing instance will refer to the vocabulary to check whether it contains keywords used in training. The corresponding TF-IDF value or one hot representation of each matched keyword will be retrieved, and construct the feature vectors for testing samples. After that, those well-trained models will be restored to predict the cluster id for testing samples. The clustering performance can be evaluated by comparing each testing sample's label and cluster ID.

4.3 Evaluation Metrics

The evaluation metrics taken in the experiment are time consumption and accuracy. The evaluation metrics taken in the experiment are time consumption and accuracy. The time consumption reflects the performance of the clustering model of handling large-scale data samples, which is the time cost from data loading to the end of clustering. The clustering accuracy is calculated using Eq. 3:

$$Accuracy = (N - e)/N \tag{3}$$

where N denotes the total number of testing samples, and e denotes the number of incorrectly clustered samples. The relation between cluster id and the label must be specified in advance to calculate the clustering accuracy. The training result can show the distribution of samples with different labels in each cluster. Hence, each cluster id is mapped to the label of the most data samples it contains.

4.4 Clustering Results and Discussion

Table 1 shows the clustering performance of three models using the extraction method of TF-IDF with a different pre-defined number of clusters. As the number of clusters rises, three models take more to predict the results. Compared with the other two models, the Spectral Clustering model takes nearly 10 times more time to complete clustering. When the scale of the data sample dramatically increases, such a gap might be more obvious.

Table 2 indicates the clustering performance of three models using the extraction method of TextRank with a pre-defined number of clusters from 3 to 8. Similarly, clustering models consume more time as the pre-defined clusters increase since more detailed clustering can take more computing steps and resources to fit each sample. However, the K-Means model with TextRank methods costs nearly three times more time than the other two models. One possible reason is that feature vectors generated by the TextRank method have high similarity, making it hard for K-Means to cluster them evenly.

Generally, clustering models with the TF-IDF method can obtain better accuracy than those with TextRank method, which means the keyword extraction method of TF-IDF is more suitable for studying short text with no strong relations between terms. The highest clustering accuracy is achieved by the SOM model with the TF-IDF extraction method when the number of pre-defined clusters is 5, that is, 64.5%. Therefore, our optimized model for the dataset utilizes SOM and TF-IDF methods

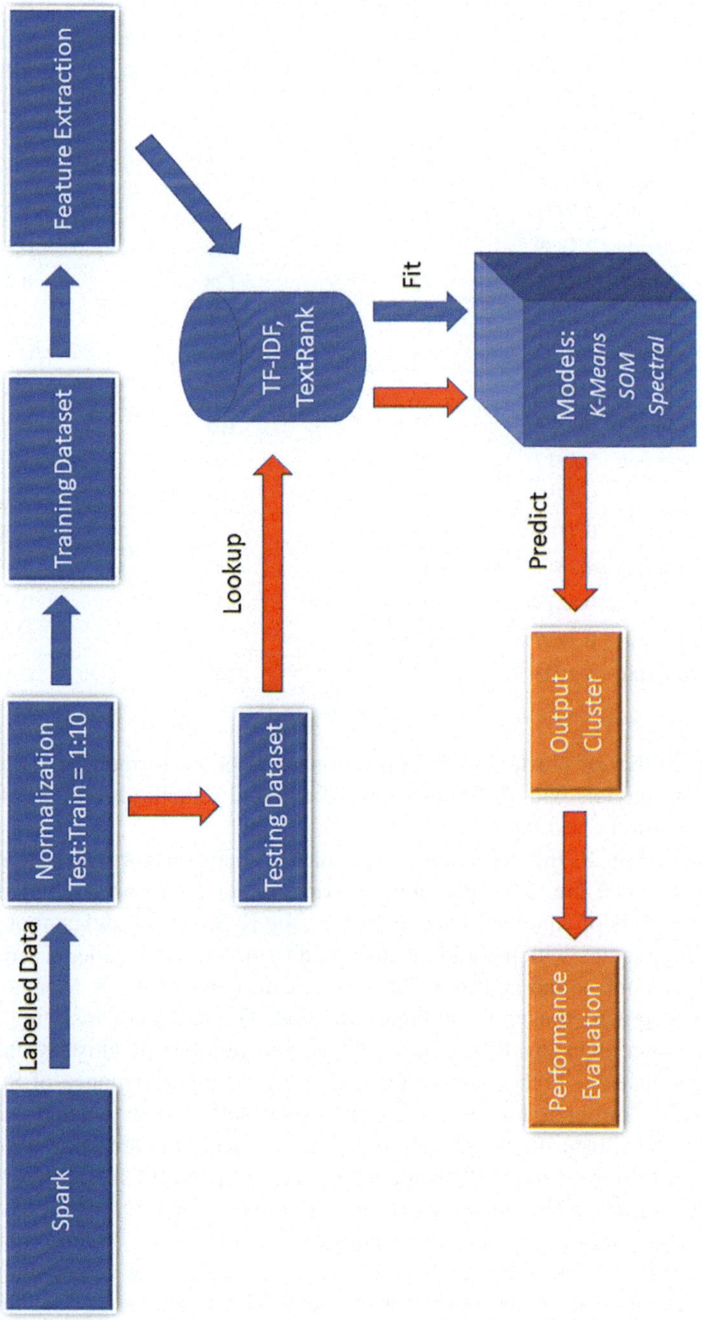

Fig. 2 Experiment process. *Source* The Authors

Table 1 Clustering results with TF-IDF extraction method

K	Cluster algorithm	Time taken (s)	Number of correctly clustered	Number of incorrectly clustered	Accuracy
3	K-means	0.3042	98	102	0.490
	SOM	0.6760	79	121	0.395
	Spectral clustering	2.5669	108	92	0.540
4	K-means	0.4348	95	105	0.475
	SOM	0.5751	74	126	0.370
	Spectral clustering	2.7839	108	92	0.540
5	K-means	0.4062	72	128	0.360
	SOM	0.6461	129	71	0.645
	Spectral clustering	4.6133	98	102	0.490
6	K-means	0.4349	74	126	0.370
	SOM	0.6539	95	105	0.475
	Spectral clustering	5.4912	109	91	0.545
7	K-means	0.4759	96	104	0.480
	SOM	0.7186	61	139	0.305
	Spectral clustering	4.2311	94	106	0.470
8	K-Means	0.5277	91	109	0.455
	SOM	0.7519	57	143	0.285
	Spectral clustering	3.2003	97	103	0.485

and sets the number of clusters to 5. Figure 3 shows the performance comparison of all clustering models with different keyword extraction methods and pre-defined cluster numbers in the testing.

As shown in Fig. 3, the K-Means model has a steady performance compared to other models since the K-Means method works under the principle of grouping samples in k clusters of equal variance. However, the K-Means model cannot obtain any outstanding results, which could be attributed to our dataset's failed assumption that clusters are convex and isotropic. These characteristics of the K-Means model make it suitable regarded as the benchmark method. The spectral Clustering model can perform better than the K-Means model as the number of clusters changes, meaning each cluster is highly non-convex. However, the performance of Spectral Clustering is less dominant than we expected. One possible reason is that the affinity matrix in Spectral Clustering is difficult to determine due to the length limitation of the data samples. Moreover, the "Discretize" option in Spectral Clustering is robust to the random initialization. However, it tends to create parcels of fairly even and geometrical shapes that might not fit our dataset.

As shown in Fig. 3, the self-organised map (SOM) accuracy with either TF-IDF or TextRank features as inputs outperforms the K-Means clustering and Spectral Clustering algorithm. Self-organized map (SOM) is renowned for its capability of

Table 2 Clustering results with TextRank extraction method

K	Cluster algorithm	Time taken (s)	Number of correctly clustered	Number of incorrectly clustered	Accuracy
3	K-means	2.1516	104	96	0.520
	SOM	0.6327	87	113	0.435
	Spectral clustering	0.9012	106	94	0.530
4	K-means	2.2373	84	116	0.420
	SOM	0.7052	66	134	0.330
	Spectral clustering	0.9023	106	94	0.530
5	K-means	2.2474	71	129	0.355
	SOM	0.7238	80	120	0.400
	Spectral clustering	0.9153	101	99	0.505
6	K-means	2.2941	54	146	0.270
	SOM	0.7448	82	118	0.410
	Spectral clustering	0.9338	121	79	0.605
7	K-means	2.6841	79	121	0.395
	SOM	0.7668	60	140	0.300
	Spectral clustering	0.9782	102	98	0.510
8	K-means	2.6210	76	124	0.380
	SOM	0.7972	45	155	0.225
	Spectral clustering	0.9928	95	105	0.475

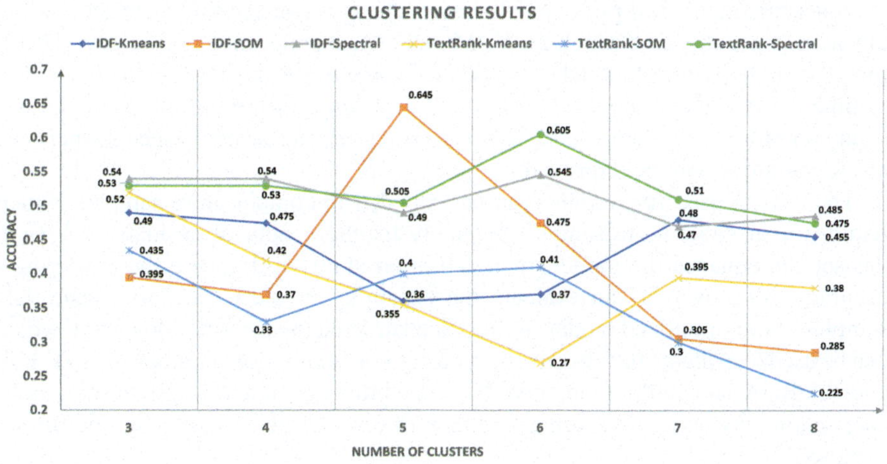

Fig. 3 Summary of the clustering results. *Source* The Authors

handling complex problems and reducing the dimensionalities into data mappings that can be easily interpreted. Therefore, SOM can process a wide range of classification problems and provide a compelling summary of the given data. The experiment dataset covers different healthcare areas, so the diversification of extracted keywords used for clustering is prominent. Such diversity can be better resolved by the SOM algorithm, generating an accurate low-dimensional interpretation of the data. However, as we can observe in Fig. 3, the results for SOM with TF-IDF show a certain degree of fluctuation. When the number of clusters equals 5, the accuracy of IDF-SOM can reach the highest among all models, but when the cluster number increases to 8, the accuracy drops to its lowest. The characteristics of the self-orgainised map (SOM) might contribute to this observation; that is, the number of pre-defined clusters has significantly impact on the clustering performance.

5 Conclusions

In this paper, the performances of three clustering models with TF-IDF and TextRank methods, respectively, have been evaluated on a real dataset from Twitter. The keyword extraction method of TF-IDF can generate a more suitable feature vector for clustering of short text than TextRank. The k-Means model shows stable clustering results that can be set as the baseline. The Spectral model with TF-IDF can achieve excellent accuracy, but its capability of clustering large-scale data samples is limited. Although the accuracy performed by the SOM model shows heavy fluctuation as the number of pre-defined clusters changes, we can find a cluster number given a dataset based on which the SOM model can achieve the best results. Therefore, the final optimized model is constructed using SOM and TF-IDF methods. Due to time constraints, part of the design and evaluation of the optimized model has been compromised to achieve basic functions. There are several aspects where the clustering model can be improved.

The dataset used for training can be extended to contain more samples, which requires us to label more samples manually for the benefit of evaluation. A large dataset can enhance the model's generalisation ability to improve the clustering's accuracy. Moreover, it can evaluate the capability of clustering large-scale data samples of our optimized model. We have constructed the dataset of long text, which can be used to evaluate the effectiveness of keyword extraction methods, i.e., TF-IDF and TextRank, in handling long text. Also, the influence of text length of the healthcare sentiments on the clustering accuracy of our optimized model can be further explored.

References

Abualigah, L., H.E. Alfar, M. Shehab, and A.M.A. Hussein. 2020. Sentiment analysis in healthcare: A brief review. In *Recent advances in NLP: The case of Arabic language. Studies in computational intelligence*, vol. 874, ed. M. Abd Elaziz, M. Al-qaness, A. Ewees and A. Dahou. Cham: Springer. https://doi.org/10.1007/978-3-030-34614-0_7.

Alguliev, R.M., and R.M. Aliguliyev. 2005. Effective summarization method of text documents. In *The 2005 IEEE/WIC/ACM international conference on web intelligence (WI'05)*, 264–271. IEEE.

Andrzejewski, D., and X. Zhu. 2009. Latent Dirichlet allocation with topic-in-set knowledge. In *Proceedings of the NAACL HLT 2009 workshop on semi-supervised learning for natural language processing*, 43–48.

Asan, U., and S. Ercan. 2012. An introduction to self-organizing maps. In *Computational intelligence systems in industrial engineering*, vol. 6, ed. C. Kahraman, 295–315. Paris: Atlantis Press.

Bahrainian, S.A., and A. Dengel. 2013. Sentiment analysis and summarization of twitter data. In *2013 IEEE 16th international conference on computational science and engineering*, 227–234. IEEE.

Belle, A., R. Thiagarajan, S. Soroushmehr, F. Navidi, D.A. Beard, and K. Najarian. 2015. Big data analytics in healthcare, *BioMed Research International* 2015.

Bracewell, D.B., F. Ren, and S. Kuriowa. 2005. Multilingual single document keyword extraction for information retrieval. In *2005 international conference on natural language processing and knowledge engineering*, 517–522. IEEE.

Brin, S., and L. Page. 1998. The anatomy of a large-scale hypertextual web search engine, computer networks and ISDN systems. In *Proceedings of the seventh international world wide web conference*, vol. 30, no. 1, 107–117 [Online]. Available at: https://www.sciencedirect.com/science/article/pii/S016975529800110X.

Chluski, A., and L. Ziora. 2015. The application of big data in the management of healthcare organizations: A review of selected practical solutions. *Informatyka Ekonomicz* (01).

Cohen, P.R., and C.A. Sutton. 2003. Very predictive ngrams for space-limited probabilistic models. In *International symposium on intelligent data analysis*, 134–142. Berlin: Springer.

Guha, S., A. Meyerson, N. Mishra, R. Motwani, and L. O'Callaghan. 2003. Clustering data streams: Theory and practice. *IEEE Transactions on Knowledge and Data Engineering* 15 (3): 515–528.

Guo, A., and T. Yang. 2016. Research and improvement of feature words weight based on tfidf algorithm. In *2016 IEEE information technology, networking, electronic and automation control conference*, 415–419.

Gupta, V.S., and S. Kohli. 2016. Twitter sentiment analysis in healthcare using hadoop and r. In *2016 3rd international conference on computing for sustainable global development (INDIACom)*, 3766–3772. IEEE.

Hanks, P. 2013. *Lexical analysis: Norms and exploitations*. Cambridge: MIT Press.

Hartigan, J.A. 1975. *Clustering algorithms*. New York: Wiley.

Hatzivassiloglou, V., and K. McKeown. 1997. Predicting the semantic orientation of adjectives. In *35th annual meeting of the association for computational linguistics and 8th conference of the European chapter of the Association for Computational Linguistics*, 174–181.

Hu, X., and B. Wu. 2006. Automatic keyword extraction using linguistic features. In *Sixth IEEE international conference on data mining-workshops (ICDMW'06)*, 19–23. IEEE.

Hu, J., S. Li, Y. Yao, L. Yu, G. Yang, and J. Hu. 2018. Patent keyword extraction algorithm based on distributed representation for patent classification. *Entropy* 20 (2): 104.

Hulth, A. 2003. Improved automatic keyword extraction given more linguistic knowledge. In *Proceedings of the 2003 conference on empirical methods in natural language processing*, 216–223.

Kaur, J., and V. Gupta. 2010. Effective approaches for extraction of keywords. *International Journal of Computer Science Issues (IJCSI)* 7 (6): 144.

Kohler, R. 2012. *Quantitative syntax analysis*, vol. 65. Walter de Gruyter.

Kohonen, T. 1982. Self-organized formation of topologically correct feature maps. *Biological Cybernetics* 43 (1): 59–69. Available at: https://doi.org/10.1007/BF00337288.

Kohonen, T. 2001. *Self-organizing maps*, vol. 30. The information sciences book series. Berlin and Heidelberg: Springer [Online]. Available at: https://doi.org/10.1007/978-3-642-56927-2.

MacQueen, J. 1967. Some methods for classification and analysis of multivariate observations. In *Proceedings of the fifth Berkeley symposium on mathematical statistics and probability*, vol. 1, no. 14, Oakland, CA, USA, 281–297.

Matsuo, Y., and M. Ishizuka. 2004. Keyword extraction from a single document using word co-occurrence statistical information. *International Journal on Artificial Intelligence Tools* 13 (01): 157–169.

Mihalcea, R., and P. Tarau. 2004. Textrank: Bringing order into text. In *Proceedings of the 2004 conference on empirical methods in natural language processing*, 404–411.

Na, S., L. Xumin, and G. Yong. 2010. Research on k-means clustering algorithm: An improved k-means clustering algorithm. In *2010 third international symposium on intelligent information technology and security informatics*, 63–67. IEEE.

Navada, A., A.N. Ansari, S. Patil, and B.A. Sonkamble. 2011. Overview of use of decision tree algorithms in machine learning. In *2011 IEEE control and system graduate research colloquium*, 37–42. IEEE.

Ohsawa, Y., N.E. Benson, and M. Yachida. 1998. Keygraph: Automatic indexing by co-occurrence graph based on building construction metaphor. In *Proceedings IEEE international forum on research and technology advances in digital libraries-ADL'98*, 12–18. IEEE.

Onoda, T., T. Yumoto, and K. Sumiya. 2008. Extracting and clustering related keywords based on history of query frequency. In *2008 second international symposium on universal communication*, 162–166. IEEE.

Ouyang, Y., W. Li, S. Li, and Q. Lu. 2011. Applying regression models to query-focused multi-document summarization. *Information Processing & Management* 47 (2): 227–237.

Pat Research. 2020. *Big data analytics and predictive analytics in 2021—Reviews, features, pricing, comparison* [Online]. Available at: https://www.predictiveanalyticstoday.com/big-data-analytics-and-predictive-analytics/.

Raghupathi, W., and V. Raghupathi. 2014. Big data analytics in healthcare: Promise and potential. *Health Information Science and Systems* 2 (1): 1–10.

Rajaraman, A., and J.D. Ullman. 2011. *Data mining*, 1–17. Cambridge: Cambridge University Press.

Salton, G., and C. Buckley. 1988. Term-weighting approaches in automatic text retrieval. *Information Processing & Management* 24 (5): 513–523.

Singh, J., G. Singh, and R. Singh. 2017. Optimization of sentiment analysis using machine learning classifiers. *Human-Centric Computing and Information Sciences* 7 (1): 32.

Su, F., and K. Markert. 2008. From words to senses: A case study of subjectivity recognition. In *Proceedings of the 22nd international conference on computational linguistics (Coling 2008)*, 825–832.

Turney, P.D. 2000. Learning algorithms for keyphrase extraction. *Information Retrieval* 2 (4): 303–336.

Uzun, Y. 2005. Keyword extraction using Naive Bayes. Bilkent University, Department of Computer Science, Turkey. Available at: www.cs.bilkent.edu.tr/guvenir/courses/CS550/Workshop/YasinUzun.pdf.

Wang, Q., W. Zhang, J. Li, F. Mai, and Z. Ma. 2022. Effect of online review sentiment on product sales: The moderating role of review credibility perception. *Computers in Human Behavior* 133: 107272. https://doi.org/10.1016/j.chb.2022.107272.

Wartena, C., and R. Brussee. 2008. Topic detection by clustering keywords. In *2008 19th international workshop on database and expert systems applications*, 54–58. IEEE.

Wodak, R. 2011. Critical linguistics and critical discourse analysis. *Discursive Pragmatics* 8: 50–70.

Zhai, Z., B. Liu, H. Xu, and P. Jia. 2011. Clustering product features for opinion mining. In *Proceedings of the fourth ACM international conference on Web search and data mining*, 347–354.

Zhang, C. 2008. Automatic keyword extraction from documents using conditional random fields. *Journal of Computational Information Systems* 4 (3): 1169–1180.

Zhu, L., A. Galstyan, J. Cheng, and K. Lerman. 2014. Tripartite graph clustering for dynamic sentiment analysis on social media. In *Proceedings of the 2014 ACM SIGMOD international conference on management of data*, 1531–1542.

Chapter 10
Big Data Analytics and the Future of Smart Transport and Healthcare Systems

Abstract Urban experimentation has long become a platform of city transitions, where we see opportunities for smart development. This applies to many sectors, including smart transportation and healthcare. In this concluding chapter, we delve into some key sectoral contributions before highlighting some of the lessons learnt. The power of Big Data analytics, through both correlational and predictive analysis, indicates various ways of facilitating knowledge for institutional development, governance enhancement, and operational optimisation. In both smart transport and smart healthcare systems, we see these transitions happening at a gradual pace. Thus, we believe these two sectors are leading the Big Data analytics research and practice, particularly in the context of smartness and smart city development. The chapter also summarises some of the key lessons from all case study chapters.

Keywords Big Data · Smart Urban Systems · Smart Cities · Transportation · Healthcare

1 A Brief Reflection on Big Data Analytics and Smart Transport and Healthcare Systems

Urban experimentation has long become a platform of city transitions, where we see opportunities for smart development (Raven et al. 2017). The two sectors we have covered in this book are critical to smart experiments, meaning they are major platforms for transformative urban development. In the study of reconciling big data for new urban science and smart city governance, Hong et al. (2022) argue that big data "*is expected to create radical new opportunities for urban research and practice*". Thus, the applications and solutions of big data-driven methods will continue to be prominent in future urban-related research, particularly in urban science, urban studies, and urbanism. The work will focus more on research formulation using big data, developing new methods for data collection and analysis, and considering data-driven outputs and knowledge representation (ibid). The so-called new urban

S. Pourroostaei Ardakani and A. Cheshmehzangi, *Big Data Analytics for Smart Transport and Healthcare Systems*, Urban Sustainability,
https://doi.org/10.1007/978-981-99-6620-2_10

science (Batty 2012, 2013; Acuto 2018; Ortman et al. 2020; Cheshmehzangi et al. 2021, 2022a) will continue to show the importance of big data in cities.

In our previous book, we thoroughly discussed the nexus between 'big data analytics' and 'smart urban systems' (Pourroostaei Ardakani and Cheshmehzangi 2023). We also reflected on the fact that "*big data technologies have underpinned new forms of urban planning and management practice represented by the rolling out of smart cities and urban laboratories*" (Hong et al. 2022). Hence, this movement to help optimise the management of urban systems and dynamics will continue to move further from theoretical grounds (Kitchin 2014) to practical solutions in smart urban systems (Pourroostaei Ardakani and Cheshmehzangi 2023). The role of data fusion (Lau et al. 2019) will become more prominent in augmenting urban innovations and those that could help enhance smartness and entrepreneurship in cities and for sustainable development (Davenport 2017; Xie et al. 2019; Kaluarachchi 2022; Cheshmehzangi et al. 2023). Thus, interdisciplinary knowledge required by urban specialists (Reese 2014) will remain an important aspect of knowledge partnerships, knowledge transfer, and empower urban sciences.

One of the areas of interest in big data analytics is the use of ICT-based and ICT-driven methods in optimising cities and city operations. A wide range of big data analytics in smart cities (Cesario 2023), in particular, shows us a range of networks and applications (Chen et al. 2014) that allow better knowledge use through deeper analysis. In this book, we have covered a large range of 'big data' applications and solutions in the research and practice of smart transport and healthcare systems. This comprehensive coverage shows the following four factors: (1) the importance of these two smart systems as part of essential or critical infrastructure in cities, (2) the role of big data-driven approaches in enhancing daily city operations, (3) data-related knowledge that requires sectoral and cross-sectoral understanding and evaluation, and (4) data pre-processing and processing that suggests responsive approaches or solutions to urban problems or challenges.

In this concluding chapter, we delve into some key sectoral contributions before highlighting some of the lessons learnt. The power of big data analytics, through both correlational and predictive analysis, indicates various ways of facilitating knowledge for institutional development, governance enhancement, and operational optimisation. The insights from the case study examples reflect on what we have covered in our previous volume, i.e., exploring performances (Albino et al. 2015), providing innovation methodologies (Almiall et al., 2012), developing big data ecosystem architecture (Anwar et al. 2021), specifying relevant software architecture for analysis (Arapakis et al. 2019), etc. With these, we also remind the readers about governance of big data (Baykurt and Raetszch 2020) and the gradual shift towards co-governance (Bifulco et al. 2017) and perhaps e-governance (Paskaleva 2009; Cheshmehzangi 2022c). In both smart transport and smart healthcare systems, we see these transitions happening at a gradual pace. Thus, we believe these two sectors are leading the big data analytics research and practice, particularly in the context of smartness and smart city development.

2 Sectoral Contributions of the Book

In the age of urban informatics (Foth et al. 2011), one can argue about the importance of business intelligence (Intellspot, n.d.). Big data analytics is evolving for nearly four to five decades (Pourroostaei Ardakani and Cheshmehzangi 2023) to ensure technologies are developed, adopted, and integrated into various urban systems. Hence, the importance of cross-sectoral big data (Laurie 2019) is visible in contemporary research, allowing us to think more about applications and solutions rather than just methodologies and directions. Previously, we touched on creating business models for smart cities (Timeus et al. 2020) to show how big data plays its part in information management and urban optimisation processes (Targio Hashem et al. 2016). In this regard, we also note the sectoral contributions of the book, particularly in two significant sectors of transport and healthcare systems. The following is a summary of chapter-by-chapter sectoral contributions of the book:

Chapter 2—A smart system is suggested to reduce the negative impacts of flight delays, allowing for better daily operations of transportation networks in cities and regions. The need for data pre-processing is evident in making a robust flight delay prediction.

Chapter 3—Traffic prediction is essential to evaluate and respond to complex problems. Thus, time-series prediction task is vital for optimising the transport data model and analysis, particularly through data pre-processing and better result accuracy.

Chapter 4—A data pattern recognition model is vital to understanding road safety and reducing transportation risks. Our work here contributed to the development of prediction models using multiple big data methods. In doing so, we proposed an approach to optimise prediction accuracy that could then be used for urban transportation system optimisation.

Chapter 5—Real-time data greatly improves mobility patterns and reduces accidents. The need for multi-spatial evaluation and multiple variables in road systems would suggest new approaches for traffic accident predictive models for better planning and prevention of risks and incidences.

Chapter 6—Investment in healthcare infrastructure development needs further analysis through the consideration of various variables. The study helps verify the use of big data to improve the availability and development of critical urban infrastructures, such as health infrastructure.

Chapter 7—In healthcare research, social media cannot be neglected. Thus, our machine learning-based approach helps verify the correction between emotion on social media and a health emergency, such as a pandemic. A prediction model suggests the importance of big data for better societal health management.

Chapter 8—Climate and environmental issues are vital to our societal health. Thus, our contribution here is to show the role of big data-enabled time series analysis for climate change analysis, such as climate change crises and impacts on communities.

Chapter 9—Similar to Chap. 7, we delve into the role of social media in achieving a healthier society but are more focused on optimised clustering model for healthcare sentiments. The applications from this chapter are broad and could help analyse larger-scale data and regions.

Lastly, the book highlights a wide range of knowledge transfers in transportation and healthcare systems. For (smart) transportation, we explicitly explored smart technologies, control and management systems, and server and tracking networks (Fig. 1). Through these, as well as various applications and solutions, we verified the importance of big data analytics processes, particularly for urban networks, mobility, and connectivity. For (smart) healthcare, apart from data-driven applications and solutions, we highlight a broad range of technologies, infrastructure, and services (Fig. 2). More importantly, this book's findings and sectoral contributions help future big data analysis research, particularly for urban management and smart urban systems. They could partly support (urban) decision-making processes, smart transitions, and urban governance (such as e-governance) processes.

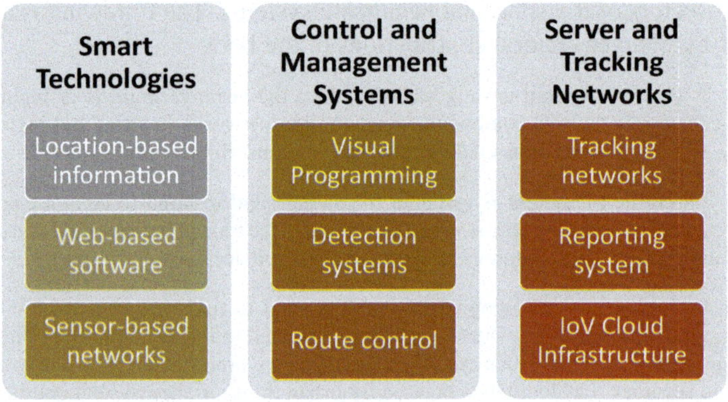

Fig. 1 Summary of smart technologies, control and management systems, and server and tracking networks for smart transportation systems in cities. *Source* The Authors

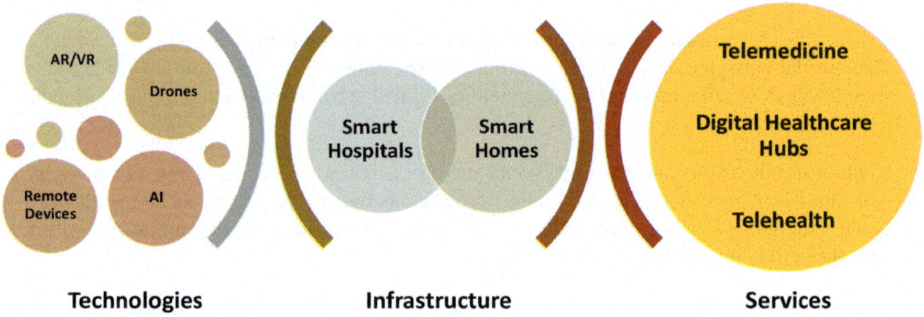

Fig. 2 The summary of 'Smart Care' in three categories of technologies, infrastructure, and services (Drawn by the authors, inspired from the hospitalist article, available from: https://www.the-hospitalist.org/hospitalist/article/242968/mixed-topics/hospital-medicine-and-future-smart-care)

3 Concluding Remarks: A Summary of Lessons Learnt for Future Research

To conclude the book, we summarise six lessons learnt for future research. For smart transport, we suggest (1) integrated transportation systems, (2) data-driven intelligence for smart mobility, and (3) the use of simulations and data-driven models for future transportation systems. For healthcare, we suggest three key areas of (1) smart technologies for smart healthcare, (2) telehealth and telemedicine, and (3) IoT-enabled smart healthcare systems. Here, we summarise these six overarching lessons learnt for future research, particularly from the perspective of big data analytics in these two sectors and other relevant sectors.

3.1 On 'Smart Transport'

- Smart City and Integrated Transportation Systems

The move to intelligent or smart transportation systems comes from a greater understanding of integration and collaborative decision-making processes. In this regard, smart cities call for a solution centered around integrating various infrastructure, stakeholders, functionalities, and services (Schaffers et al. 2011; Ruchira and Sengupta 2023). According to Visan et al. (2022), the operations need a "*joint effort of standardization, interconnection and assimilation of emerging technologies to meet operational, safety and environmental objectives at metropolitan, national or international level*". This means the need for a sort of mobility ecosystem, functional architecture, and collaborative network is essential in developing the capacities and capabilities of smart transportation systems. In this regard, big data technologies would be able to help optimise the networks and systems that require better collaboration in processes and operations (Pourroostaei Ardakani et al. 2023).

- Data-Driven Intelligence for Smart Mobility

Building resilience into smart mobility has become an important way of optimising smart urban transportation systems (Mageto et al. 2022). In practice, we see a growing demand for data-driven intelligence for mobility and transportation (Bosch, n.d.), while academic research is still behind some of these directions. The value-added information, as suggested by Bosch (ibid), could be seen as a suitable way to cover "*the entire vehicle life cycle*…[which]…*can be leveraged to generate better products, shorter development times, and additional data-driven services*". Some of their other suggested features include (1) customer digital twin for various uses and within the data-driven intelligence framework, (2) modular architecture for mix-and-match solutions, (3) safe and securing handling of data, (4) information generation and transfer, (5) services aligned with data-driven intelligence and systems, and (6) customer services and integration (ibid). Thus, the model suggests data transfer into the cloud, using machine learning and AI algorithms, semantic models, and domain-specific rules, and considers the life-cycle and user-independent data for the platform. Such an approach would then enable multi-modal and data-driven platforms for future smart mobility to move away from vehicles alone, i.e., towards larger-scale of logistics, smart mobility applications, smart investment, network growth, etc.

- Learning from Simulations and Data-Driven Models

As Osaba et al. (2020) suggested, we expect to see more approaches to data-driven optimisation for transportation and smart mobility applications. The ICT-based methods are expected to help enhance communication infrastructures as well as optimise daily operations, management of resources, and governance of systems. Thus, in the contemporary age, data-driven modelling and learning, particularly in science and engineering (Montáns et al. 2019) is developing rapidly. Various examples of predicting traffic and road accidents show the importance of integrating data-driven and simulation models in urban networks and analysis (Shafiei et al. 2022; Cheshmehzangi and Pourroostaei Ardakani 2021; Pourroostaei Ardakani et al. 2022). Such approaches help optimise traffic management, reduce accidents, have a better incident impact analysis, and create data-driven platforms for future uses.

3.2 On 'Smart Healthcare'

- Smart Technologies for Smart Healthcare

Apart from the many data-driven approaches we highlighted in the book, it is important to highlight the role of ICTs, IoT, mobile internet, cloud computing, AI, etc., in various sectors of smart healthcare (Tian et al. 2019). Smart technologies are used "*to transform the traditional medical system in an all-round way, making healthcare more efficient, more convenient, and more personalised*" (ibid). Five key smart technologies are proposed by Impactmybiz (2023), including (1) Remote Monitoring and Automated Healthcare Systems, (2) Telehealth and Remote Care, (3) Connected Emergency Response Solutions, (4) Smart Hospital Management, and (5) AI in Healthcare. In sum, the so-called game-changing approach in the healthcare industry cannot be completed without the use of innovative new tools, data-driven approaches, and new technologies.

- Telehealth and Telemedicine Era

The growing demand for digital healthcare indicates using big data in various ways, particularly in telehealth and telemedicine. According to Agrawal and Prabakaran (2020), big data "*will be an integral part of the next generation of technological developments—allowing us to gain new insights from the vast quantities of data being produced by modern life. There is significant potential for the application of Big Data to healthcare, but there are still some impediments to overcome, such as fragmentation, high costs, and questions around data ownership. Envisioning a future role for Big Data within the digital healthcare context means balancing the benefits of improving patient outcomes with the potential pitfalls of increasing physician burnout due to poor implementation leading to added complexity*". In this regard, we expect to see shifts from digital healthcare to telehealth platforms, where transactions, interactions, data, and processes are digital and optimised. This would be considered the age of big tech and big data for new data-driven opportunities and technologies, such as blockchain, and move toward developing health data marketplaces, consumerisation of healthcare (Thomason 2021), and digital platforms.

- IoT-Enabled Smart Healthcare Systems

Associated with some of our earlier work (Pourroostaei Ardakani and Cheshmehzangi 2023), we refer to the convergence of smart city with IoT and big data (Singh and Kumar 2022), to be even more progressive in smart healthcare systems (Pourroostaei Ardakani 2021; Zhao et al. 2021). Termed as IoT-Health (Internet-of-Things in Healthcare), we see a large potential "*to create a pervasive environment for monitoring patient health and safety as well as improving how physicians deliver care. It can also boost patient engagement and satisfaction by allowing them to spend more time in the comfort of their home and interacting with their care centers whenever needed. A significant driver for the IoT-Health market is the increasing penetration of connected devices in healthcare*" (NIST 2022). Thus, there are many examples of novel smart healthcare systems, services, and applications (Rani et al. 2022) that suggest a diverse array of "*cutting-edge technologies in modern healthcare delivery*". This emerging role indicates the shift beyond just digital platforms and toward integrated data-driven models and networks.

To conclude, we reflect back on the book's aim and objectives, demonstrating the sectoral contributions of the book for future research and practice in (smart) healthcare and transport systems. This big data analytics era will have its opportunities and challenges, and we have to absorb and utilise what we progress in a timely manner. Big data analytics will continue to grow, and it will be up to us to decide how it can be used better and humanely. Still, many research gaps and challenges need to be addressed to ensure game-changing methods, applications, and solutions are viable and user-friendly. We must understand the growing issues of the digital divide and its barriers and consider how we may need to break them rather than make new ones. While we focused on sector-based studies, we also note the need to break barriers between sectors, allowing for transformative urban transitions to happen in a healthy—and hopefully—sustainable way. The pathways in which transport and healthcare systems develop indicate new opportunities for ICT-based and data-driven applications, if not mere methods and directions, to achieve smart urban systems. Thus, we again urge researchers, practitioners, and governments to develop novel ideas and directions through partnership capacities. We have to shift away from our current and increasing over-reliance on technologies and focus more on how technologies and big data are utilised to optimise and enhance our livelihoods. Making places better is not far from our reach; thus, we have to use every opportunity to overcome any urban challenges and pressures we may have in the coming years or decades. To advance some of these directions, we have to create humane pathways and innovative solutions for the future of cities.

References

Acuto, M. 2018. Global science for city policy. *Science* 359 (6372): 165–166. https://doi.org/10.1126/science.aao2728.

Agrawal, R., and S. Prabakaran. 2020. Big data in digital healthcare: Lessons learnt and recommendations for general practice. *Heredity* 124: 525–534. https://doi.org/10.1038/s41437-020-0303-2.

Albino, V., U. Berardi, and R.M. Dangelico. 2015. Smart cities: Definitions, dimensions, performance, and initiatives. *Journal of Urban Technology* 22 (1): 3–21.

Anwar, M.J., A.Q. Gill, F.K. Hussain, and M. Imran. 2021. Secure big data ecosystem architecture: Challenges and solutions. *EURASIP Journal on Wireless Communications and Networking*, Article Number 130. https://doi.org/10.1186/s13638-021-01996-2

Arapakis, I., Y. Becerra, O. Boehm, G. Bravos, V. Chatzigiannakis, C. Cugansco, G. Demetriou, I. Eleftheriou, J.E. Mascolo, L. Fodor, S. Ioannidi, D. Jakovetic, L. Kallipolitis, E. Kavakli, D. Kopanaki, N. Kourtellis, M.M. Marcos, R.M. de Pozuelo, N. Milosevic, G. Morandi, E.P. Montanera, G. Ristow, R. Sakellariou, R. Sirvent, S. Skrbic, I. Spais, G. Vasiliadis, and M. Vinov. 2019. Towards specification of a software architecture for cross-sectoral big data applications. In 2019 IEEE World Congress on Services (SERVICES). Conference Held in Milan, Italy. https://doi.org/10.1109/SERVICES.2019.00120

Batty, M. 2012. Building a science of cities. *Cities* 29 (Supplement 1): S9–S16. https://doi.org/10.1016/j.cities.2011.11.008.

Batty, M. 2013. Big data, smart cities and city planning. *Dialogues in Human Geography* 3(3). https://doi.org/10.1177/2043820613513390

Baykurt, B., and C. Raetszch. 2020. What smartness does in the smart city: From vision to policy. *Convergence: The International Journal of Research into New Media Technologies* 26(4). https://doi.org/10.1177/1354856520913405

Bifulco, F., M. Tregua, and C.C. Amitrano. 2017. Co-governing smart cities through living labs. Top evidences from EU. *Transylvanian Review of Administrative Sciences* 50: 21–37.

Bosch. n.d. *Data-driven intelligence: Unlocking vehicle data potential.* Available from: https://www.bosch-mobility.com/en/solutions/software-and-services/data-driven-intelligence/

Cesario, E. 2023. Big data analytics and smart cities: applications, challenges, and opportunities. In *Frontiers in big data*, Vol. 6, Section on Data Mining and Management. https://doi.org/10.3389/fdata.2023.1149402

Chen, M., S. Mao, and Y. Liu. 2014. Big data: A survey. *Mobile Networks and Applications* 19: 171–209. https://doi.org/10.1007/s11036-013-0489-0.

Cheshmehzangi, A. 2022c. ICT for a well-functioning government in cities: Towards integrated information platforms for E-Government and E-Transformation. In *ICT, cities, and reaching positive peace*, 47–67. Singapore: Springer.

Cheshmehzangi, A., and S. Pourroostaei Ardakani. 2021. Urban traffic optimization based on modeling analysis of sector-based time variable: The case of simulated Ningbo, China. *Front. Sustain. Cities* 3. https://doi.org/10.3389/frsc.2021.629940

Cheshmehzangi, A., A. Dawodu, and A. Sharifi. 2021. *Sustainable urbanism in China.* New York: Routledge.

Cheshmehzangi, A., Y. Li, H. Li, S. Zhang, X. Huang, X. Chen, Z. Su, M. Sedrez, and A. Dawodu. 2022a. A hierarchical study for urban statistical indicators on the prevalence of COVID-19 in Chinese city clusters based on multiple linear regression (MLR) and polynomial best subset regression (PBSR) analysis. *Scientific Reports* 12: 1964. https://doi.org/10.1038/s41598-022-05859-8.

Cheshmehzangi, A., Z. Su, and T. Zou. 2023. ICT applications and the COVID-19 pandemic: Impacts on the individual's digital data, digital privacy, and data protection. In *Frontiers in human dynamics*, Section on Digital Impacts, Vol. 5. Available from: https://doi.org/10.3389/fhumd.2023.971504

Davenport, T.H. 2017. How analytics has changed in the last 10 years (and how it's stayed the same). In *Harvard business review: Analytics services*. Available from: https://hbr.org/2017/06/how-analytics-has-changed-in-the-last-10-years-and-how-its-stayed-the-same.

Foth, M., J.H.J. Choi, C. Satchell. 2011. Urban informatics. In *Proceedings of the ACM 2011 Conference on Computer Supported Cooperative Work*, Hangzhou, China, 1–8, 19–23 Mar 2011.

Hong, A., L. Baker, R.P. Curierl, J. Duminy, B. Buswala, C. Guan, and D. Ravindranath. 2022. Reconciling big data and thick data to advance the new urban science and smart city governance. *Journal of Urban Affairs*. https://doi.org/10.1080/07352166.2021.2021085.

Impactmybiz. 2023. *5 examples of smart technology in healthcare*. Available from: https://www.impactmybiz.com/blog/smart-technology-in-healthcare/.

Intellspot. n.d. *Big data technologies: List, stack, and ecosystem in demand*, part of Business Intelligence. Available from: https://www.intellspot.com/big-data-technologies/?ref=quuu.

Kaluarachchi, Y. 2022. Implementing data-driven smart city applications for future cities. *Smart Cities* 5(2):455–474. https://doi.org/10.3390/smartcities5020025

Kitchin, R. 2014. *The data revolution: Big data, open data, data infrastructures and their consequences*. Thousand Oaks: Sage Publications.

Lau, B.P.L., S.H. Marakkalage, Y. Zhou, N. Ul Hassan, C. Yuen, M. Zhang, and U.-X. Tan. 2019. A survey of data fusion in smart city applications. *Information Fusion* 52: 357–374. https://doi.org/10.1016/j.inffus.2019.05.004.

Laurie, G.T. 2019. Cross-sectoral big data. *ABR* 11: 327–339. https://doi.org/10.1007/s41649-019-00093-3.

Mageto, J., H. Twinomurinzi, R. Luke, S. Mhlongo, K. Bwalya, and S. Bvuma. 2022. Building resilience into smart mobility for urban cities: An emerging economy perspective. *International Journal of Production Research*. https://doi.org/10.1080/00207543.2022.2139866.

Montáns, F.J., F. Chinesta, R. Gomez-Bombarelli, and J.N. Kutz. 2019. Data-driven modeling and learning in science and engineering. *Comptes Rendus Mécanique* 347 (11): 845–855. https://doi.org/10.1016/j.crme.2019.11.009.

NIST. 2022. *Internet-of-Things in Healthcare (IoT-Health)*, Project/programme under the United Stated Government. Available from: https://www.nist.gov/programs-projects/internet-things-healthcare-iot-health.

Ortman, S.G., J. Lobo, M.E. Smith, and P.F. Biehl. 2020. Cities: Complexity, theory and history. *PLoS ONE* 15 (12): e0243621. https://doi.org/10.1371/journal.pone.0243621.

Osaba, E., J.J. Sanchez Medina, E.I. Vlahogianni, X.-S. Yang, A.D. Masegosa, J. Perez Rastelli, and J. Del Ser. 2020. Data-driven optimization for transportation logistics and smart mobility applications. In *IEEE intelligent transportation systems magazine*, vol. 12, no. 4, 6–9. https://doi.org/10.1109/MITS.2020.3017033.

Paskaleva, K.A. 2009. Enabling the smart city: The progress of city E-governance in Europe. *International Journal of Innovation and Regional Development* 1 (4): 405–422. https://doi.org/10.1504/IJIRD.2009.02273.

Pourroostaei Ardakani, S., and A. Cheshmehzangi. 2023. *Big data analytics for smart urban systems*. Singapore: Springer. In press.

Pourroostaei Ardakani, S., X. Lian, K.T. Mengistu, R.S. So, X. Wei, B. He, and A. Cheshmehzangi. 2023. Road car accident prediction using a machine-learning-enabled data analysis. *Sustainability* 15(7):5939. https://doi.org/10.3390/su15075939.

Pourroostaei Ardakani, S., T. Xia, A. Cheshmehzangi, and Z. Zhang. 2022. An urban-level prediction of lockdown measures impact on the prevalence of the COVID-19 pandemic. *Genus* 78(1):28. https://doi.org/10.1186/s41118-022-00174-6.

Pourroostaei Ardakani, S., X. Wu, X. Pan, and X. Gao. 2021. ARM: A real-time health monitoring mobile application. In *5th EAI International Conference, COMPSE 2021*, July 29. https://doi.org/10.1007/978-3-030-87495-7_4.

Rani, S., M. Rajagopal, N. Kumar, S.H. Ahmed Shah. 2022. *IoT-enabled smart healthcare systems, services and applications*. New York: John Wiley & Sons, Inc. https://doi.org/10.1002/9781119816829.

Raven, R., F. Sengers, P. Spaeth, L. Xie, A. Cheshmehzangi, and M. de Jong. 2017. Urban experimentation and institutional arrangements. *European Planning Studies* 27(2). Urban Experimentation & Sustainability Transitions. https://doi.org/10.1080/09654313.2017.1393047.

Reese, L. 2014. The present and future of urban affairs research. *Journal of Urban Affairs* 36 (sup2): 543–550. https://doi.org/10.1111/juaf.12143.

Ruchira, G., and D. Sengupta. 2023. Smart urban metabolism: A big-data and machine learning perspective. In *Urban metabolism and climate change: Perspectives for sustainable cities*, eds. R. Bhadouria, S. Tripathi, P. Singh, P.K. Joshi, and R. Singh, 325–344. Cham: Springer. https://doi.org/10.1007/978-3-031-29422-8_16.

Schaffers, H., N. Komninos, M. Pallot, B. Trousse, M. Nilsson, and A. Oliveira. 2011. Smart cities and the future internet: Towards cooperation frameworks for open innovation. In *The future internet*, ed. J. Domingue, A. Galis, A. Gavras, T. Zahariadis, D. Lambert, F. Cleary, and M. Nilsson, Vol. 6656, 431–446. Heidelberg: Springer. https://doi.org/10.1007/978-3-642-20898-0_31.

Shafiei, S., A.-S. Mihaita, H. Nguyen, and C. Cai. 2022. Integrating data-driven and simulation models to predict traffic state affected by road incidents. *Transportation Letters: The International Journal of Transportation Research* 14(6). https://doi.org/10.1080/19427867.2021.1916284.

Singh, A., and A. Kumar. 2022. *Convergence of smart city with IoT and Big Data*. Available from: https://sciforum.net/paper/view/13930.

Targio Hashem, I.A., V. Chang, N.B. Anuar, K. Adewole, I. Yaqoob, A. Gani, E. Ahmed, and H. Chiroma. 2016. The role of big data in smart city. *International Journal of Information Management* 36 (5): 748–758. https://doi.org/10.1016/j.ijinfomgt.2016.05.002.

Thomason, J. 2021. Big tech, big data and the new world of digital health. *Global Health Journal* 5 (4): 165–168. https://doi.org/10.1016/j.glohj.2021.11.003.

Tian, S., W. Yang, J.M. Le Grange, P. Wang, W. Huang, and Z. Ye. 2019. Smart healthcare: Making medical care more intelligent. *Global Health Journal* 3 (3): 62–65. https://doi.org/10.1016/j.glohj.2019.07.001.

Timeus, K., J. Vinaixa, and F. Pardo-Bosch. 2020. Creating business models for smart cities: a practical framework. In *Public management review*, vol. 22, issue 5. Special issue: Management, Governance and Accountability for Smart Cities and Communities. https://doi.org/10.1080/14719037.2020.1718187.

Visan, M., S.L. Negrea, and F. Mone. 2022. Towards intelligent public transport systems in smart cities; Collaborative decisions to be made. *Procedia Computer Science* 199: 1221–1228. https://doi.org/10.1016/j.procs.2022.01.155.

Xie, L., A. Cheshmehzangi, M. Tan-Mullins, A. Flynn, and T. Heath. 2019. Urban entrepreneurialism and sustainable development: A comparative analysis of Chinese eco-developments. *Journal of Urban Technology* 27(1). https://doi.org/10.1080/10630732.2019.1680940.

Zhao, P., P. Kar, and S. Pourroostaei Ardakani. 2021. ANTON: Activity recognition-based smart home control system. *SN Computer Science* 2:428. https://doi.org/10.1007/s42979-021-00824-0.

Index

A
Abnormal data process, 130
Accident reduction, 12, 97
Accuracy, 11, 14, 32, 33, 45–52, 56, 59, 67, 68, 73, 75, 78, 82–84, 93, 110, 115, 127, 128, 130–132, 142, 144, 145, 150–153, 157, 158, 166, 168–170
AI algorithms, 144, 180
AIC, 83
Air Traffic Management (ATM), 32
Algorithm, 8, 10, 29, 48, 52, 67, 72, 73, 83, 84, 93, 97, 107, 109, 110, 124, 126, 130, 143, 158, 161–163, 168–170
Analytic Hierarchy Process (AHP), 83
Apache Spark, 33, 49, 65, 68
API, 124, 125, 165
AR, 178
Artificial Intelligence (AI), 3, 4, 8, 124, 144, 180
Artificial Intelligence of Things (AIoT), 144
Artificial Neural Network (ANN), 13, 33, 68, 83, 106, 141, 146, 148, 149, 153, 154, 162
Auto-generated data, 159
Automated healthcare systems, 180
Automobile monitoring system, 67

B
Backpropagation, 143, 145, 146
Bayesian network solution, 82
Belgium, 11, 45, 46, 48, 51, 53–55, 57, 59
Benchmark, 11, 33, 47, 63, 65, 74–78, 83, 109, 127, 128, 168
BERT model, 125
Bidirectional, 51
Big data analysis techniques, 64
Big Data Analytics (BDA), 2–10, 30, 78, 175–179, 181
Big data technology, 124, 176, 179
Big data visualisation, 32
Billing, 159
Binary classification, 66, 151
Biometric data, 159
Blockchain, 181
Boolean value, 84, 86
Brazil, 13, 32, 141, 142, 150, 154
Brussels, 46, 48
Build predictive model, 68
Business intelligence, 30, 177
Business model, 177

C
CAN-BUS, 67
Canonical Interval Forest (CIF), 47
Car-trip, 65
Catch22, 47
China, 13, 30, 64, 83, 121–126, 131, 134, 135, 138
Classification, 9, 13, 47, 48, 59, 65–68, 73, 75, 77, 78, 110, 112, 124, 132, 141, 145, 148, 151, 154, 160, 162, 170
Classifiers, 10, 11, 47, 63, 65–68, 76, 77, 82, 83, 93, 109, 110, 146
Climate change, 13, 15, 141, 142, 144, 146, 148, 150, 152, 154, 156, 177
Climate dataset, 142

S. Pourroostaei Ardakani and A. Cheshmehzangi, *Big Data Analytics for Smart Transport and Healthcare Systems*, Urban Sustainability,
https://doi.org/10.1007/978-981-99-6620-2

Clustering, 9, 13, 14, 67, 68, 73, 78, 83, 157, 158, 160–166, 168–170, 177
Clustering algorithms, 158, 163, 168
Coefficient, 63, 71, 76, 78, 111, 112, 136, 146
Complexity, 2, 7, 82, 110, 143, 145, 181
Complex structure, 159
Computing technology, 144
Consumerisation, 181
Contextual, 124
Convenient, 3, 16, 115, 131, 180
Convergence, 153, 181
Convolutional Neural Networks (CNN), 48, 145, 165
Correlation calculation, 82, 86, 93
Correlation matrix, 147
Corresponding numeric label, 85
Country road network, 46
COVID-19, 12, 13, 21, 53, 64, 103–107, 110, 117, 121–126, 129, 131–137
COVID-19 information, 122, 123, 136
CRISS, 66
Cross-sectoral, 176, 177
Cross-validation, 37, 50, 110, 112, 113, 142–144, 151, 154
CSV, 49, 56, 84
Current Expenditures on health Per capita (CEP), 107, 110, 111
Current health expenditure Per capita, PPP (CPPP), 107, 111

D

Database, 1–3, 8, 20, 31, 84, 159
Data cleaning, 31, 49, 59, 83, 108
Data-driven, 2, 3, 9, 11, 30, 65, 68, 78, 106, 175, 176, 178, 180, 181
Data-driven intelligence, 179
Data-driven model, 107, 179–181
Data-driven platform, 180
Data extraction, 31
DataFrame, 48, 49, 56, 59
Data integration, 31
Data normalization, 158, 165
Data ownership, 8, 180
Data privacy, 7
Data quality, 7, 8, 31, 37, 78, 146
Data reduction, 31
Data resampling, 68, 72
Data scientist, 31
Data transformation, 31, 49, 146
DayOfWeek, 50
Decentralised, 5
Decision tree, 11, 12, 33, 63, 65, 67, 68, 75–77, 82, 83, 103, 110, 116, 117, 132, 137, 142, 145, 151, 153, 154, 161
Deep learning framework, 67
Departure, 33, 35–38, 51
Detection system, 4
Device, 6, 8, 15, 20, 24, 71, 132, 181
Dew point temperature, 141, 142
DGT, 83
Digital healthcare, 180, 181
Digital healthcare hubs, 178
Digital platform, 181
Dimension reduction, 32, 142, 146
Directorate on traffic accidents, 83
Discretise data samples, 65
Dissatisfaction, 32
DNB, 33
Domain-specific rules, 180
Down-sampling, 72, 150
Driver's rush, 11, 63, 71, 72, 76, 77
Driver's well-being, 11
Driving pattern, 11, 63–68, 73, 74, 77, 78
Driving pattern clustering, 74
Driving speed, 66
Drones, 15
DTANN, 83
Dynamic sentiment analysis, 162
Dynamic Time Warping (DTW), 37, 47

E

E-commerce, 14, 157, 158
Economic benefit, 141
E-governance, 176, 178
Emotion analysis, 13, 121, 126, 127, 130
Empowerment, 13, 121
Energy consumption, 5, 18
Engineering, 2, 46, 49, 56, 59, 64, 180
Environmental conditions, 66, 68
Epi-center, 122
Epidemiological dynamics, 106
Ethiopian road traffic, 82
Euclidean distance, 110, 132, 163
Europe, 32
Experimentation, 175
Extreme value, 42, 116

F

Facebook, 125, 157, 159
Feature extraction algorithm, 162
Feed-forward artificial neural network, 146
Flight status, 10, 29

Fragmentation, 180
Framework, 2–4, 22, 33, 104, 105, 124, 158, 165, 179
Fraud detection, 4, 159
Freight transport, 11, 45, 46, 59
Frequency, 2, 32, 47, 124, 131, 133, 160–162
Fully Convolutional Network (FCN), 11, 45, 48, 52, 56, 58, 59
Fuzzy C means clustering algorithm, 162

G
Game-changing, 180, 181
Gaussian Naive Bayes model, 93
Gaussian Parsimonious Bayesian model, 82
GDP, 106, 107, 117
General Government Health Expenditure (GGHE), 105
Generation Z, 126
Global Health Security (GHS), 105
Governance enhancement, 176
Granada highway, 83

H
Hadoop, 3
Hadoop distributed file system, 48
Healthcare industry, 6, 180
Healthcare infrastructure, 12, 103–106, 111, 117, 177
Healthcare investment, 12, 103, 105–107, 110, 117
Healthcare sentiment, 13, 158, 170, 177
Health data marketplace, 181
Healthy Life Expectancy (HALE), 105
Heat map, 36, 69
Hidden Markov model, 67
High-quality data, 49
HIVE-COTE, 47, 48
HowNet, 127
HTML tag, 161
Human factor, 7
Hunan University, 125
Hybrid data resampling, 11, 63

I
ICT-based, 176, 180, 181
ICT-driven, 9, 176
ICU, 105
Inception-v4, 48
InceptionTime, 47, 48
Inequalities, 103
Infection, 6, 107
INMET, 142
Innovation, 3, 4, 176
Innovative, 180, 182
Instagram, 125, 157
Integrated transportation systems, 179
Intensive care unit beds, 105
Interconnection, 179
International health regulation scores, 105
Internet of Things (IoT), 8, 18, 144, 179–181
Interoperability, 7
IoT-enabled, 181
IoT-health, 181
IoV Cloud infrastructure, 178
Italian, 124

K
Keras, 51
Key-value pair, 85
Keyword classification, 161
Keywords extraction, 160, 161
KM, 74
K-means algorithm, 74
K-means clustering, 162, 163, 168
K-mode clustering, 83
K-Nearest Neighbours algorithm (KNN), 11, 12, 33, 63, 65, 68, 78, 83, 103, 104, 109, 110, 115–117, 145
Knowledge discovery, 33
Knowledge inference, 159

L
Latent Dirichlet allocation, 124
Lateral acceleration, 72
LAX, 32
LCC, 83
Least Squares Support Vector Machine (LSSVM), 107
Life-cycle, 179, 180
Light freezing drizzle, 84
Light rain, 84
Linear regression, 12, 33, 38, 40, 42, 52, 84, 103, 104, 107, 109, 110, 116, 117, 132, 137
Linguistics approaches, 160
Liveable cities, 1, 9, 10
Location-based information, 178
Lockdown, 122, 123, 126
Logistics, 6, 30, 65, 68, 75–77, 180

Long Short-Term Memory (LSTM), 11, 32, 45, 46, 48, 50–52, 56, 58–60, 144–147, 154

M

Machine learning (ML), 4, 10, 11, 13, 29, 31–33, 37, 40, 45–47, 64–68, 71, 76–78, 83, 84, 107, 116, 121, 127, 129, 131, 132, 137, 142, 145, 147, 151, 154, 161–163, 177, 180
Machine learning approaches, 11, 32, 65, 161
Machine learning-enabled, 9, 10, 29, 32
Maximum entropy, 127
Maximum relevance, 12, 81, 82, 84, 86
Mean Absolute Error (MAE), 51, 52, 56, 132, 137, 144
Mean Absolute Percentage Error (MAPE), 132, 137, 144
MEASEL, 47
Metropolitan, 37, 179
Microblogs, 128, 130
Midwife, 105
Min-Max normalisation, 108, 117
MLlib, 65, 147, 148
MOBIlity-Advanced Indicators Dashboard (MOBI-AID), 46
Model generalisation, 65
Model parameters, 56, 109, 153
Model performance, 37, 50, 56, 86, 111
Model training, 33, 65, 67, 68, 72, 73, 78, 84, 104, 111, 113, 151, 153, 165
Model training delay, 65, 67, 77, 78
MR algorithm, 82, 93
Multi-class classification, 66, 148
Multi-Layer Perception (MLP), 11, 63, 65, 76, 77, 145, 146, 150, 151
Multilayer Perceptron Classifier, 148
Multilayer Reception, 68
Multinomial Logistics Regression, 11, 63, 65, 68, 75–77, 83
Multi-objective Info-Fuzzy Network (M-IFN), 67
Multi-target classification, 65–67, 72
Multi-target prediction, 11, 65–68, 72, 73
Multi-target prediction model, 11, 63, 65, 67, 68, 72, 77, 78
Multivariate Time Series Classification (MTSC), 47, 48
MUSE, 47
Mutual information, 12, 81, 82, 86, 87, 93–95, 127

N

Naïve Bayes, 33, 68, 82, 83, 127, 161
Network growth, 180
Nigeria, 105, 107, 126
Ningbo, 64
NoSQL, 3, 8

O

Obstacle status, 66
Onboard Diagnostics (OBD), 68
On-board Unit (OBU), 11, 45, 46, 48, 50, 59
On-site visiting experience, 158
Optimisation, 12, 31, 46, 49, 60, 63, 64, 93, 97, 115, 176, 177, 180
Optimised classification, 10

P

Pandas, 48, 49
Pandemic prevention, 12, 103, 105, 106
Parking period, 66
Partitioned clustering algorithm, 162
Passenger count, 71, 76
Patterns, 2–5, 8, 12, 30–33, 37, 46, 47, 53, 59, 64, 68, 78, 81, 83, 84, 97, 143, 145, 159, 177
Pearson's Correlation Coefficient, 11
Pearson technique, 71
Poisson regression models, 84
Polynomial regression, 10, 29, 30, 37, 38, 40–42
Population density, 91
Prediction accuracy, 11, 12, 32, 33, 47, 59, 63, 68, 143, 153, 177
Prediction results, 11, 37, 38, 42, 51, 56, 57, 59, 63, 65, 68, 73, 74, 93
Predictive model, 12, 150, 177
Principal Component Analysis (PCA), 142, 143, 145, 146, 150, 153, 154
Probability, 21, 72, 90, 160
Production management, 141
Psychological model, 126
Purchasing decision, 158
PVG, 32
PySpark, 33, 40, 48, 50, 56, 131, 147, 148, 154
Python web crawler, 128
PyTorch, 147, 154

Q

Qingbo, 124

Quantile discretisation, 11, 63

R
Rain intensity, 11, 63, 71, 72, 75, 76
Random convolutional kernel transform, 47
Random Forest (RF), 11, 33, 47, 63, 65, 67, 68, 72, 73, 75–78, 83, 132, 137, 145, 154
Random Forest Regression and Maximal Information Coefficient (RFR-MIC), 33
Real-time data, 6, 8, 12, 81, 177
Regional transportation, 64
Remote devices, 178
Remote monitoring, 180
Replication, 72
Reporting system, 15
Resilience, 106, 179
Resilient, xii
ResNet, 48
Result accuracy, 11, 45, 56, 177
Road orientation, 83
Road separation, 83
Road Side Units (RSU), 46
Road status, 11, 63, 68
Road width, 66
Robust statistics, 51
RobustScaler, 50
ROCKET, 47
Root Mean Squared Error (RMSE), 52, 56, 57, 132, 137, 144
Route control, 178
RPM, 64, 71, 75
R-Squared value, 46

S
Safety hazard, 88
Scalability, 8, 47
Security, 4, 5, 7, 8, 10, 16, 159
Self-Organized Map (SOM), 14, 157, 158, 162, 163, 165, 166, 168–170
Semantic models, 180
Sensor data, 159
Sensor-based networks, 178
Sentiment, 13, 14, 121, 124–127, 129, 130, 137, 157–162
Sentimental analysis, 123–125
Shanghai, 32, 134
Simulation models, 180
SIR model, 107
Smart homes, 10
Smart hospital management, 3, 180
Smart hospitals, 3, 178, 180
Smart investment, 180
Smart mobility, 179, 180
Smart mobility applications, 180
Smartness, 176
Smart technologies, 3, 15, 18, 178–180
SO-PMI algorithm, 127
Social event, 14, 157, 158
Social media analysis, 13
Social networks, 13, 121–123, 131, 157, 160
Societal impact, 13, 121
Spark, 48, 49, 56, 59, 123, 131, 137, 138, 158, 165
Spark MLlib, 147
SparkSQL, 11, 45, 46, 48, 49, 56
Spatial-temporal, 8
Spectral Clustering, 162, 164–166, 168
Speed limit, 64
Stakeholder, 1, 7, 179
Stock market, 31
Street environment factors, 64
Street function, 64
S-type interconnection, 145
Sub-city, 83
Sum of squared errors, 73
Support Vector Regression (SVR), 12, 33, 103, 109, 112, 113, 117
Surveillance systems, 4, 32
Sustainable development, 105, 176

T
TapNet, 48, 52
Telehealth, 21–24, 179–181
Telemedicine, 24, 179, 180
TensorFlow, 51, 147, 154
Term Frequency-Inverse Document Frequency (TF-IDF), 124, 160, 162
Testing Index, 106
TextBlob, 124–126
TextRank, 158, 160, 162, 165, 166, 168, 170
Time sequence dataset, 50
Time sequence prediction, 50
Time series, 11, 13, 45–50, 53, 59, 67,, 142, 144, 145, 147, 154, 177
Time series model, 142, 144, 146, 149
Timestamp, 11, 45, 46, 49, 50, 53, 56, 59
Topic detection, 161
Tracking networks, 178
Traffic, 2, 6, 9–12, 16, 18, 32, 45–48, 50, 53, 59, 63, 64, 66, 78, 81–84, 88, 93, 144, 145, 177, 180

Traffic accident, 12, 81–84, 86, 88, 93, 177
Training data, 50, 52, 73, 106, 109, 110, 127, 132, 160, 163
Transactional data, 159
Transportation network, 12, 29, 46, 64, 81, 97, 177
Transportation network systems, 60
Travel behaviour factors, 64
TributaryPCA, 143
Tropical infectious diseases, 105
Truck, 46, 49, 53–55
Twitter, 13, 14, 124–126, 157–159, 165, 170

U
Union, 49
United Nations (UN), 81, 105
United States, 12, 30, 32, 81, 82, 126
Universal Health Coverage (UHC), 105
Up-sampling, 72
Urban challenge, 181
Urban flow analysis, 6
Urban management, 178
Urban planning, 1, 6–9, 176
Urban road traffic accident analysis, 84
Urban transition, 181
US accidents dataset, 12, 81, 82
US airlines, 10, 29, 30
US economy, 30
User-independent, 180
Uttarakhand, 83

V
Validation, 143, 165
Variety, 2, 4, 30–32, 65, 126
Vector, 32, 47, 66, 67, 72–74, 78, 107, 109, 132, 161, 163, 165, 166, 170
Vehicle, 11, 19, 45, 46, 59, 63–68, 71, 72, 75, 76, 83, 84, 179, 180
Vehicle factors, 64, 68
Velocity, 2, 4, 31, 32
Veracity, 31, 32
Visualisation, 31, 32, 46, 53, 131, 137, 160, 162
Visual programming, 178
Volume, 2, 4–6, 8, 9, 31, 32, 53, 71, 176
Voting, 67, 73
VR, 178

W
Weather data, 141, 144–146
Weather factors, 64
Web-based software, 178
Web data, 159
WeChat, 138
Weibo, 13, 121–130, 132, 133, 137, 138
Weka, 82, 83
Word segmentation, 130, 131, 163
World Bank, 81, 104
World Health Organisation (WHO), 8, 81, 104

X
X-axis, 131

The manufacturer's authorised representative in the EU is Springer Nature Customer Service Centre GmbH, Europaplatz 3, 69115 Heidelberg, Germany. If you have any concerns regarding our products, please contact ProductSafety@springernature.com

Printed and bound by CPI Group (UK) Ltd, Croydon, CR0 4YY
01/10/2026
02225877-0017